DRUG ACTIONS ON CHOLINERGIC SYSTEMS

Pharmacology Monographs.
Series editor : P. B. Bradley

DRUG ACTIONS ON CHOLINERGIC SYSTEMS

R. W. Brimblecombe,

B.Sc., M.Sc., Ph.D., D.Sc., F.I.Biol.

Head of the Pharmacology Group and Deputy Director,
The Research Institute,
Smith Kline & French Laboratories Ltd.,
Welwyn Garden City.

M

First published 1974 by
THE MACMILLAN PRESS LTD.
London and Basingstoke
Associated companies in New York Dublin
Melbourne Johannesburg and Madras

SBN 333 17636 7

Printed in Great Britain by
Thomson Litho, East Kilbride, Scotland

Preface

This monograph would be more appropriately titled "Some Aspects of Drug Actions on Cholinergic Systems". The contents reflect the particular interests and views of the author. This has allowed these aspects of the subject to be discussed in detail.

The original brief given to the author was to concentrate mainly on the central nervous system; some explanation therefore is required for the fact that nearly one-fifth of this volume deals with drug actions on peripheral systems. The part of the book in question is concerned, in the main, with a consideration of the interactions between drugs and receptors and is meant to serve as a prelude to the rest of the book which explores the hypothesis that central effects of various drugs result from their interactions with receptors which seem to be analogous to those involved in drug actions at peripheral cholinoceptive sites. It seems essential to obtain information concerning the structural requirements for activity at these peripheral receptors using techniques which give reasonably precise results before attempting any considerations of central actions where the available techniques give results which are, on the whole, far from exact.

Thus, the early chapters contain considerations of the actions of drugs at postganglionic parasympathetic endings ("muscarinic" receptors) and at autonomic ganglia and neuromuscular junctions ("nicotinic" receptors). Since there have been suggestions that the ganglionic receptors are in many ways analogous to some central acetylcholine receptors the question of whether differences exist between the ganglionic and the neuromuscular junction receptors is considered. Additionally, a critical examination is made of the claim that some drugs act presynaptically at the ganglion to cause a release of acetylcholine which, in turn, interacts with receptors on the post-synaptic membrane. If such a mechanism operates peripherally the possibility of there being a similar central system must be considered. In fact the conclusion is reached that the concensus of opinion is not in favour of this presynaptic drug action.

Anticholinesterase agents are considered with particular emphasis on drugs which antagonise their effects. Again both peripheral and central actions are discussed. The two are so closely related that it seems sensible to consider them together. A particular section of this chapter is, however, devoted to behavioural effects of anticholinesterases, that is, specifically central actions.

Two chapters of the monograph are devoted to drug actions at postulated central muscarinic and nicotinic receptors. These actions are discussed in some detail especially in terms of their effects on the behaviour of the organism as a whole. Relatively little attention is paid to drug actions at the neuronal level since this aspect of the subject has been discussed in detail by others.

My former colleagues at C.D.E. Porton with whom a good deal of the work mentioned in this monograph was carried out include Drs. Bryan Barrass, Alan Bebbington and Tom Inch who have with great patience tried to instil a little fundamental chemical knowledge into my biologist's brain, although they should not be held responsible for any errors or naive statements, and Drs. David Buxton and David Green, my fellow pharmacologists. I am grateful to them all, for much helpful discussion and many of the ideas propounded here have, in fact, emanated from them.

Finally, my sincere thanks are due to Mrs. Pauline Smith and Miss Hilda Churcher from C.D.E. Porton and to Mrs. Madeline Oliver from Smith Kline & French who are responsible for typing various parts of the manuscript.

Foreword

This book introduces a new series of Pharmacology Monographs, which will be published at approximately yearly intervals, and which will cover various aspects of pharmacology, with particular reference to the pharmacology of the central nervous system. Each monograph will be principally concerned with a single topic or area of current pharmacological interest, and the author, or authors, will be experts in that field. Apart from reviewing the subject, although not necessarily comprehensively, as the monographs are not intended to be reviews, they may present personal or speculative views. It is hoped that the books will have an appeal not only to pharmacologists, including postgraduate students and research staff, but also to research workers in other, related disciplines, including biochemistry, physiology and experimental psychology, and to some clinical specialities, for example, anaesthetics, neurology and psychiatry, depending upon the particular topic of the monograph. Thus, although primarily pharmacological, the books will have an interdisciplinary appeal.

The choice of Cholinergic Systems as the topic for the first monograph is appropriate, as this is an area of pharmacological research which has been somewhat neglected in recent years, in favour of the more popular biogenic amines, the catecholamines and tryptamines. However, there is evidence that interest in cholinergic mechanisms is reviving, at least as far as the central nervous system is concerned, particularly in relation to drug action. Thus, the advent of many new centrally acting drugs, directly or indirectly involved with cholinergic transmission has stimulated this interest. On the other hand, we are still lacking much precise information about the actions of acetylcholine at synapses in the central nervous system, particularly as regards the presence of 'nicotinic' and 'muscarinic' receptors in the brain, whereas much more information is available about peripheral cholinergic mechanisms. Thus, Dr. Brimblecombe has chosen to review the evidence from the periphery in some detail, and to use this data to discuss possible analogies between peripheral and central acetylcholine receptors. Whilst not everyone may agree with this approach, its very discussion may lead to further investigations which should help to elucidate the role of cholinergic mechanisms in the brain and possibly stimulate further the development of new drugs.

P. B. Bradley,
Birmingham,
July 1974.

Contents

PREFACE		v
FOREWORD		vii
CHAPTER 1.	Historical Introduction	1
	References	13
CHAPTER 2.	Drug Actions at Peripheral Muscarinic Sites	19
	2.1 Agonists	19
	2.2 Antagonists	30
	References	39
CHAPTER 3.	Drug Actions at Peripheral Nicotinic Sites	43
	3.1 Actions of some ganglion-stimulant drugs	45
	3.2 Neuromuscular effects of some antimuscarinic drugs	47
	3.3 Ganglionic and neuromuscular junction receptors	50
	3.4 The nature of nicotinic receptors	54
	References	58
CHAPTER 4.	Anticholinesterase agents	63
	4.1 Acute pharmacological effects of anticholinesterases	66
	4.1.1 Muscarinic actions	66
	4.1.2 Nicotinic actions	67
	4.1.3 Central actions	67
	4.1.4 Effects resulting from actions other than the inhibition of cholinesterase	68
	4.2 Chronic effects of anticholinesterases	69
	4.3 Treatment of anticholinesterase poisoning	71
	4.3.1 Protection of the enzyme from the inhibitor	72
	4.3.2 Reactivation of the enzyme	76
	4.3.3 Antagonism of the effects of acetylcholine	87
	(i) Ganglion-blocking drugs	
	(ii) Curarising agents	
	(iii) Antiacetylcholine drugs	
	4.3.4 Symptomatic treatment and miscellaneous drugs	93
	4.3.5 Combinations of drugs	97
	4.4 Behavioural effects of anticholinesterases	98
	4.4.1 Experimental animals	98
	4.4.2 Man	111
	References	121

CHAPTER 5.	Drug Actions at Central Muscarinic Sites	133
	5.1 Evidence for central muscarinic receptors	133
	5.2 The involvement of central cholinergic systems in the control of movement	138
	5.3 Central actions of muscarinic agonists	142
	5.4 Central actions of antiacetylcholine drugs	145
	5.4.1 Mode of action	145
	5.4.2 Effects on man	148
	5.4.3 Effects on electrical activity of the brain	152
	5.4.4 Effects on animal behaviour	161
	(i) Observational methods	
	(ii) Social behaviour	
	(iii) Spontaneous motor activity	
	(iv) Emotional behaviour	
	(v) Spontaneous alternation	
	(vi) Conditioned behaviour	
	5.5 The role of central muscarinic mechanisms in behaviour	177
	References	
CHAPTER 6.	Drug Actions at Central Nicotinic Sites	181
	6.1 Evidence for central nicotinic receptors	191
	6.2 Central actions of nicotine and related	191
	substances	194
	6.2.1 General effects	
	6.2.2 Behavioural effects	194
	(i) Spontaneous activity	198
	(ii) Conditioned behaviour	
	6.2.3. Neurophysiological effects	
	6.2.4 General conclusions	205
	References	210
CHAPTER 7.	General Conclusions.	211
		215

1 Historical Introduction

The synthesis of acetylcholine was first reported by Baeyer in 1867 but its physiological role was not recognised for a number of years. In 1907 Dixon noted similarities between the effects of muscarine and of stimulating the vagus nerve and he suggested that the latter effected a release of a muscarine-like substance which combined with some constituent of the end-organ, in this case the heart, to manifest the effects of vagal stimulation. This suggestion was met with some ridicule and Dixon seems not to have pursued the matter. A year previously Hunt and Taveau (1906) had described the great potency of acetylcholine as a depressor agent. Then in 1914 Dale published his classic paper on "The action of certain esters and ethers of choline and their relation to muscarine". This remarkable piece of work has formed the basis for many subsequent pharmacological studies of the actions of acetylcholine-like and acetylcholine-antagonising drugs at various sites in the peripheral nervous system. Dale found that acetylcholine and some other esters and ethers of choline had a depressor action at low doses; higher doses produced a reduction in heart rate. Section of the vagi had no influence on these cardiovascular effects of acetylcholine, which were thus assumed to be peripheral in origin. Atropine completely abolished the effects but if larger doses of acetylcholine were administered to the atropinised animal then an increase in blood pressure was seen. This effect, also peripheral since it was not abolished by destruction of the bulbar centres or indeed of the whole spinal cord, could be blocked if a dose of nicotine was given, sufficient to paralyse all the autonomic ganglia. Dale described the depressor, cardio-inhibitor action of acetylcholine as being of the muscarine type since it was very similar to that brought about by the latter alkaloid. The pressor action, on the other hand, was said to be of the nicotine type. Actions of acetylcholine on other organs and tissues were examined in detail and it was concluded that the muscarine action mimicked very closely the effects of stimulation of para-sympathetic nerves while the nicotine action was recognised as being, to some extent, analogous to that of curare as well as involving the autonomic ganglia. From this pioneering work of Dale it has become conventional to refer to the actions of acetylcholine and related drugs at post-ganglionic parasympathetic endings as being "muscarinic" and those at the autonomic ganglia and at the skeletal neuromuscular junction as being "nicotinic". It is worth noting, however, that Dale himself never used these particular adjectives. Dale also, incidentally, noted the extreme liability of acetylcholine to hydrolytic decomposition and suggested the presence of an esterase capable of splitting acetylcholine into the pharmacologically relatively inert choline and acetic acid.

No historical account of acetylcholine, however brief, would be complete without reference to the brilliant work of Otto Loewi who, in 1921, published the first real proof of the role of this substance as a chemical neurotransmitter. He noted that a perfusate of the stimulated vagus nerve of a frog caused a slowing of the rate of heart beat of a second frog. The substance in the perfusate responsible for this effect he named *Vagusstoff* or vagus-substance. Five years later in 1926 Loewi and Navratil identified *Vagusstoff* as being almost certainly acetylcholine. Experiments similar to those performed by Loewi in 1921 were later carried out on mammals, notably by Rylant (1927) using rabbits and by Feldberg and Krayer (1933) using cats and dogs. All reached a similar conclusion, that acetylcholine was released at vagal endings as a neurotransmitter substance.

According to the work of Dale and Loewi it was clear that many drugs exerted their actions by mimicking or competing with acetylcholine at its various sites of action in the nervous system and soon a good deal of knowledge was amassed concerning these drugs. The next important milestone in the story was reached in 1933 when another classic series of experiments was reported by A. J. Clark. These experiments paved the way for a whole new field of study in pharmacology, that of the nature of the interaction between drugs and receptors. Clark's work is as well known as that of Dale and Loewi but, in a monograph which will be so concerned with receptors as this one, it would be improper not to give a brief account of his experiments and to consider the implications of the results.

Clark made use of the property of acetylcholine reported earlier by Dale and others, that of producing a reduction in heart rate. In experiments using the isolated frog's heart he found the concentration of the drug which reduced the rate by half. In fact, a dose of 0.02 μg was required for this effect and by calculating the number of molecules of acetylcholine involved and measuring the total available cell surface area, Clark found that these molecules would only occupy about 0.016% of the cell surface area. Similar results were obtained on suitable isolated tissues with ouabain, atropine, adrenaline and histamine and the implication was that these drugs interacted with specific receptive areas on the cell surface. These areas have become known as receptors, a term coined by Langley in 1878, and a concept used by Ehrlich at the beginning of the twentieth century, in his work with potential chemotherapeutic drugs.

Clark assumed that the interaction of a drug with a receptor was a process analogous to the adsorption of a molecule on the surface of a catalyst and he suggested that the activity of a drug, whether as an agonist or an antagonist, was a function of its affinity constant or its adsorbability. Later it became apparent that this concept could not explain all the experimental facts and Ariens (1954) and Stephenson (1956) both suggested that agonists had an additional property which they termed intrinsic activity or efficacy respectively, and which enabled the drug to produce an effect. Thus, in the context of this monograph, acetylcholine and all cholinergic drugs possess both affinity and intrinsic activity or efficacy, whereas antagonists of acetylcholine or of other agonists acting at cholinergic receptors possess only affinity. For more detailed consideration of this topic, and for the basic mathematics involved, reference should be made to Barlow (1964).

Mention should also be made at this juncture of the two main theories of drug receptor interaction. Most early workers assumed that drugs exerted their actions for the period of their occupation of the receptor—the occupancy theory. In 1961, however, Paton suggested that the rate of combination of the drug with the receptor was an important parameter—the rate theory. Agonists would have a high rate-constant of association with the receptors while antagonists would have a relatively low rate-constant. The differences between these two theories remain unresolved but, in any case, the controversy is not directly relevant to the majority of the material in this monograph.

So in the early 1930s acetylcholine was accepted as being the neurotransmitter at postganglionic parasympathetic endings, at the autonomic ganglia and at the skeletal neuromuscular junction; with this knowledge the various actions of drugs at these cholinoceptive sites began to be understood. At all sites where neurohumoral transmission is involved five types of drug action are possible:

(1) Interference with transmitter synthesis or release
(2) Potentiation of transmitter release
(3) Interference with transmitter destruction
(4) Agonist action mimicking the action of the transmitter
(5) Antagonist action blocking the action of the transmitter or exogenously administered agonist drug

Drugs which possess, or which have been claimed to show, each of these types of action will be considered in lesser or greater detail in this monograph. In this introductory chapter mention will only be made of pioneering work with drugs of each type.

Botulinus toxin is a substance which prevents the release of acetylcholine in all cholinergic fibres which have been studied. Its specificity for cholinergic systems was used by early workers, notably Edmunds and Long (1923) and Dickson and Shevky (1923), to map the distribution of cholinergic nerve fibres. These authors thought, mistakenly, that the action of the toxin was analogous to that of curare on skeletal, and of atropine on smooth muscle. It was not until the pharmacology of the toxin was re-examined by Ambache in 1949 that it became apparent that its action was at the nerve endings in preventing transmitter release. Subsequently it has become apparent that the toxin blocks conduction at the pre-terminal portion of the axon (Koelle, 1965a) but its precise mode of action is still not known.

Hemicholinium (HC-3; see formula 1) was described by Schueler in 1955. It reduces the rate of acetylcholine release from all cholinergic nerve endings by interfering with its synthesis (MacIntosh, Birks and Sastry, 1956). Specifically it blocks the system whereby choline, the metabolic precursor of acetylcholine, is concentrated in the nerve terminals (Birks and MacIntosh, 1961).

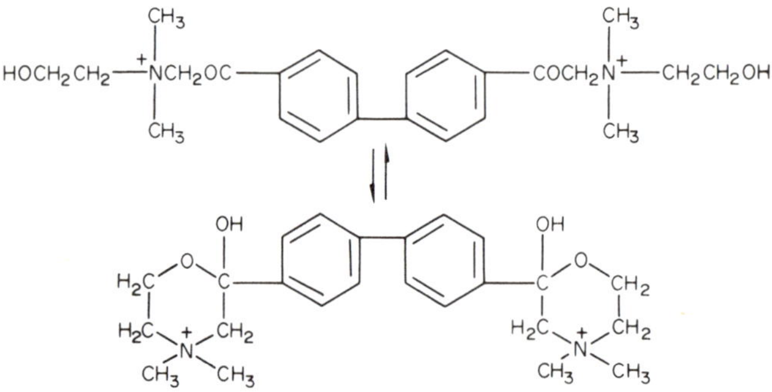

1

There seem to be drugs which specifically promote the release of acetylcholine from nerve terminals. This action has been proposed to explain, in part, the ganglion stimulant action of carbachol (Volle and Koelle, 1965). This matter will be discussed later in the monograph and evidence will be presented which seems to refute the suggestion. It has also been stated (Koelle, 1970a) that tetraethylammonium (TEA), generally regarded as a ganglion blocker, potentiates acetylcholine release from certain nerve terminals.

An important class of drugs, the anticholinesterases, interfere with the process of transmitter destruction by inhibiting the enzyme acetylcholinesterase which catalyzes the hydrolysis of acetylcholine to choline and acetic acid. Although of relatively small importance clinically, drugs of this class have been widely used as tools in investigations to elucidate various aspects of cholinergic transmission. Broadly speaking, anticholinesterases can be divided into reversible and irreversible inhibitors of the enzyme. The former, mainly carbamates, interact with the enzyme at its anionic and esteratic sites, to form a carbamylated enzyme. This, although it is more resistant to hydrolysis than the acetylated enzyme formed in the reaction between acetylcholine and the enzyme, still undergoes spontaneous hydrolysis relatively rapidly to release the free enzyme. The so-called irreversible inhibitors, mainly organophosphorus compounds, interact with the enzyme only at the esteratic site to form a phosphorylated enzyme which is very resistant to hydrolysis so that the enzyme remains inhibited for long periods.

Historically, physostigmine (see formula 2) was the first reversible inhibitor to be studied. It is a constituent of the calabar or ordeal bean, the fruit of *Physostigma venenosum* which was used in ritual trials in Africa. The bean was broken up and given to the accused to swallow. Those who vomited survived the ordeal and were considered to be innocent while the guilty who did not vomit died. Physostigmine was first isolated from the bean in 1864 by Jobst and Hesse but it was not until 1930 that Engelhart and Loewi, and then Matthes, demonstrated that its action was that of an inhibitor of acetylcholinesterase. Earlier, in 1925, Stedman and Barger had elucidated the chemical structure of physostigmine and since that time various other compounds structurally related to physostigmine have been prepared. Some, including neostigmine, pyridostigmine and ambenonium have been used clinically in the treatment of myasthenia gravis and glaucoma.

The first reported synthesis of an irreversible cholinesterase inhibitor of the organophosphorus type was in 1854 when Clermont described the preparation of tetraethylpyrophosphate (TEPP; see formula 3) but it was not until 1943 that interest in this type of compound became more general. In that year Lange and Krueger described the synthesis of dimethyl and diethyl phosphorofluoridates and also mentioned some of their physiological effects, notably their action in causing tightness of the chest and blurred vision. Later, compounds of this type were developed by Schrader (1952) as insecticides and, during World War II, some organophosphorus compounds were developed and stockpiled by the Germans as potential chemical warfare agents—the nerve agents.

Much fuller accounts of the historical developments of the anticholinesterase agents have been given by Holmstedt (1963), Koelle (1970b) and Karczmar (1970).

Following the work of Dale (1914) it has become increasingly apparent that many drugs owe their pharmacological effects to mimicking the actions of acetylcholine at various cholinergic sites, that is by interacting

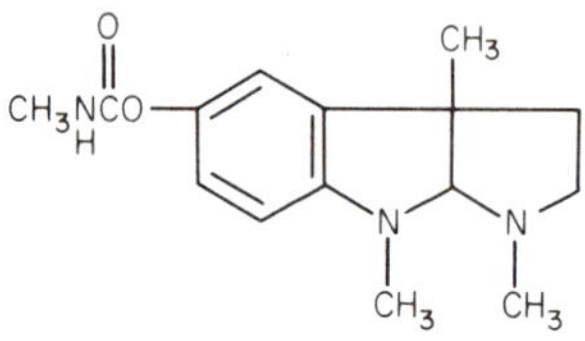

2

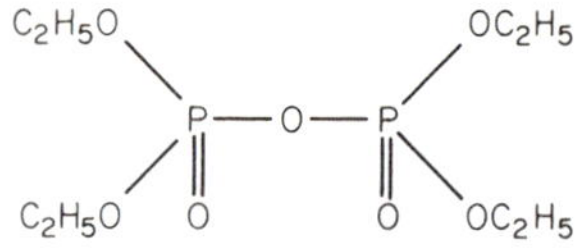

3

with acetylcholine receptors. Possibly the earliest recognition of a drug which has acetylcholine-like actions at parasympathetic endings was by Schmiedeberg and Koppe (1869) using muscarine (see formula 4) although of course, they did not recognise the full significance of their findings. Over the years much effort has been devoted to elucidating structure-activity relationships among muscarinic drugs and specific aspects of this subject will be dealt with in Chapter 2 of this monograph. A comprehensive account of the chemical pharmacology of these drugs is given by Barlow (1964).

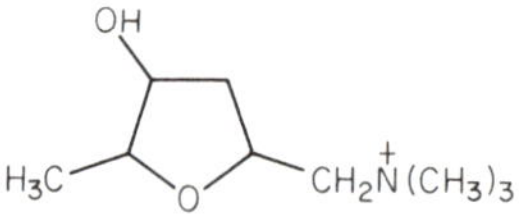

4

The actions of nicotine (see formula 5) on autonomic ganglia were investigated by Langley and Dickinson in 1889 but earlier than that in 1868 Crum Brown and Fraser had recognised the ganglionic actions of synthetic onium compounds. Again, for a complete review of the field, reference should be made to Barlow (1964) or to Volle and Koelle (1970).

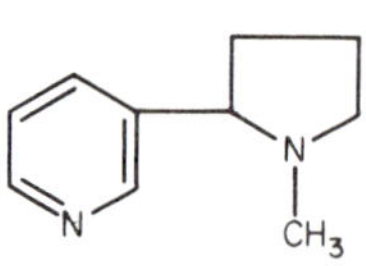

5

The recognition by Dale (1914) that small doses of nicotine imitated the actions of acetylcholine at the skeletal neuromuscular junction has already been referred to. Much earlier than this Claude Bernard (1856, 1857) and his co-workers had shown that the paralytic effects of curare were due to blocking action at the neuromuscular junction. Again, the experiments were so fundamental as to warrant brief mention here. Essentially the procedure was to inject curare into a decapitated frog, one hind leg of which was ligatured. Whereas this ligatured limb showed a withdrawal response when immersed in dilute acid, the unligatured hind limb showed no such response. Since the nerves to both limbs were equally exposed to the drug and the muscle on the unligatured side was capable of responding to direct electrical stimulation, it was assumed, correctly, that the curare was acting as a blocker at the neuromuscular junction. The structure of d-tubocurarine (see formula 6), one of the main active principles of curare, was established by King in 1935. Since then various synthetic drugs with curare-like action have been prepared, notably by Bovet and his colleagues (see Bovet, Bovet-Nitti and Marini-Bettolo, 1959) who first reported gallamine (see formula 7), for example.

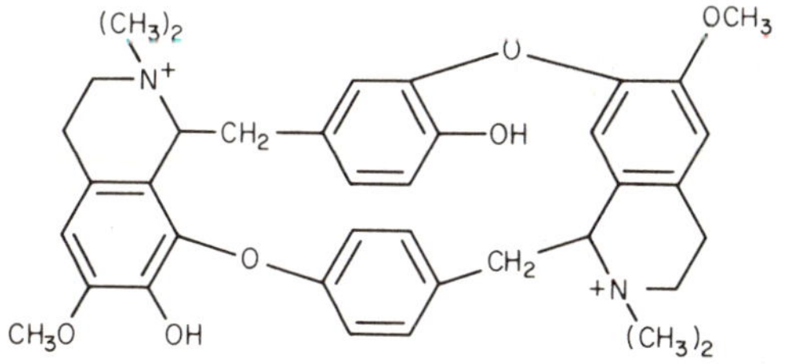

6

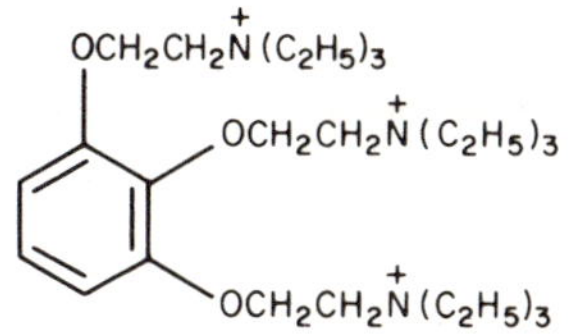

7

It soon became apparent in studies of drugs interacting with acetylcholine receptors at the neuromuscular junction that the situation was a complicated one and it was not really feasible to consider stimulant and blocking drugs separately. The actions of true curare-like drugs, described as competitive or non-depolarising neuromuscular blocking agents, are well understood but other drugs like nicotine and indeed the natural transmitter, acetylcholine itself have differing actions depending on their concentration at the active site. Low doses stimulate, whereas higher doses produce a depolarisation of the motor end-plate and the adjoining muscle cell membrane giving a neuromuscular blockade. There are also drugs like decamethonium, (see formula 8 n = 10), (Barlow and Ing, 1948; Paton and Zaimis, 1949) classed as depolarizing blocking agents which, as their name suggests, produce an immediate depolarisation at the end-plate.

$$(CH_3)_3\overset{+}{N}(CH_2)_n\overset{+}{N}(CH_3)_3$$

8

Another point worth mentioning here, although it will be considered in more detail subsequently, is that drugs differ in their relative potencies in producing blockade at autonomic ganglia and at the neuromuscular junction, both described as nicotinic sites of action of acetylcholine. Thus hexamethonium (see formula 8, n = 6) is much less potent than decamethonium as a neuromuscular agent although it is a very effective ganglion blocker. Much fuller accounts of drugs which act at the neuromuscular junction are given by Barlow (1964) and Koelle (1970c).

Drugs which antagonise the actions of acetylcholine at postganglionic parasympathetic endings have been known for many centuries in the form of extracts from *Atropa belladonna* and various other solanaceous plants. Atropine (see formula 9) was first isolated in the first half of the nineteenth century and soon its ability to block the effects of parasympathetic stimulation were demonstrated (Innes and Nickerson, 1965). As early as 1864, Kleinwachter described the use of atropine for treating poisoning with physostigmine in man.

Other naturally occurring alkaloids, notably hyoscine (see formula 10) soon became known and subsequently large numbers of semi-synthetic and synthetic drugs with atropine-like actions (formerly known as anticholinergic or antimuscarinic but now generally referred to as antiacetylcholinic drugs) have been prepared. Some of these drugs will be discussed in considerably more detail later in this monograph. A general account of their chemical pharmacology is to be found in Barlow (1964).

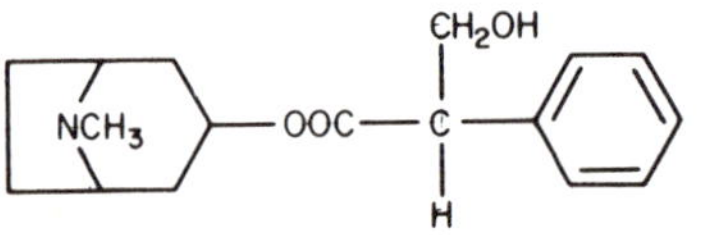

9

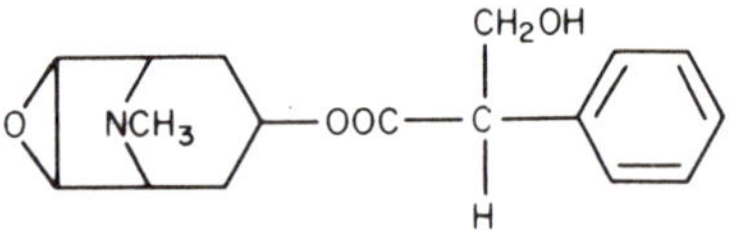

10

The biphasic actions of nicotine at the neuromuscular junction, with low doses producing stimulation and higher doses blockade, have already been referred to. Nicotine and various other drugs act similarly at the autonomic ganglia. There are, however, also a series of drugs which produce a non-depolarising type of ganglion-blockade, presumably analogous to the action of curare at the neuromuscular junction. Indeed, curare itself produces ganglion blockade but the dose required is about ten times higher than that required to produce neuromuscular paralysis. The first synthetic ganglion blocker studied seems to have been tetraethylammonium (TEA; see formula 11). Burn and Dale (1915) showed that this drug antagonised the nicotine-like actions of acetylcholine at autonomic ganglia. In contrast, the closely related tetramethylammonium (TMA; see formula 12) proved to be a ganglion stimulant.

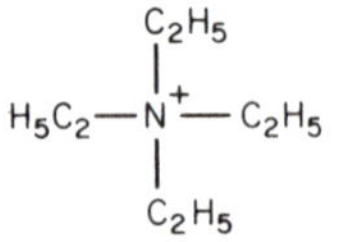

11

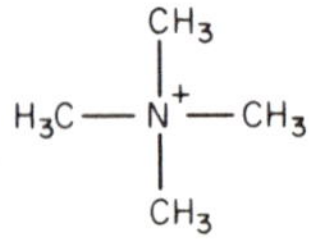

12

Subsequently, probably the next really significant advance in this field was made by Paton and Zaimis (1949, 1951) in their work on a series of bis-onium salts. They found particularly, that pentamethonium (see formula 8, n = 5) and hexamethonium (see formula 8, n = 6) were potent non-depolarising ganglion blockers.

These drugs, being quaternary salts, are, however, poorly absorbed when taken orally and so interest developed in secondary or tertiary bases and two of the earliest drugs of this type were mecamylamine (see formula 13) (Stone, Torchiana, Navarro and Beyer, 1956) and pempidine (see formula 14) (Corn and Edge, 1958; Spinks, Young, Farrington and Dunlop, 1958).

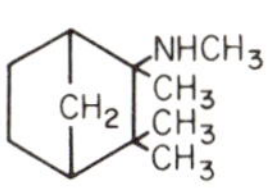

13

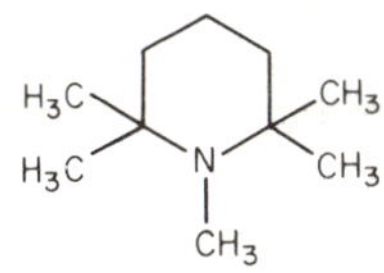

14

This introductory chapter, so far, has been concerned with the actions of drugs on peripheral cholinergic systems. A good deal of this monograph will be concerned with central actions of drugs and so it is necessary to give an equally brief introductory historical account of this aspect of the subject. Again, no attempt will be made to give a comprehensive review; rather, some of the more important milestones will be referred to.

That acetylcholine is a chemical transmitter substance in the central nervous system seems to be beyond all reasonable doubt. The evidence was summarised by Bradley (1968). When drug actions on central cholinergic systems are considered, one of the fundamental questions to be answered is whether the central acetylcholine receptors are directly analogous to those in the periphery. The first direct evidence concerning this was obtained by Curtis and Eccles (1958). Using the technique of applying drugs to single neurones by iontophoretic administration, they studied the Renshaw cells in the anterior horn of the spinal cord. These cells are internuncial neurones which synapse with collaterals from motoneurones and with other motoneurones and, being inhibitory in function, prevent the spread of excitability from one group of motoneurones to another which is not involved in any particular motor function. From the work of Curtis and Eccles it became apparent that acetylcholine is the neurotransmitter at the synapse between the collaterals of motor axons and the Renshaw cells and, more particularly, that this was a nicotinic site of action of acetylcholine. Its effects could be mimicked by nicotine and blocked by nicotinic blocking drugs.

All cholinoceptive neurones in the central nervous system are not, however, nicotinic (Bradley, 1968). Some respond rather specifically to muscarinic agents, others respond differentially to nicotinic and muscarinic agents, and it has been conjectured that there may be two types of receptors on some single neurones. From such evidence it seems reasonable to expect that drugs with various types of actions at peripheral cholinergic sites might be expected to show central effects providing they can gain access to central neurones. It is generally accepted that quaternary ammonium salts cannot penetrate the so-called "blood-brain barrier" and so would not be expected to have central actions unless they are injected directly into the ventricles or the solid matter of the brain. Such substances will not be considered in any detail in the section of this monograph dealing with the central nervous system. Attention will be focused in the main on those drugs which manifest their central effects when administered systematically.

Possibly the earliest demonstrations of the central effects of muscarinic agonists occurred in cases of poisoning with the mushroom *Amanita muscaria*. The symptoms, described by Koelle (1970d), include vertigo, confusion, coma and convulsions, all effects presumably originating in the central nervous system and due, apparently, to muscarine. It has also been suggested (Fabing, 1956) that the hallucinations which result from the ingestion of certain species of *Amanita* are also due to the muscarine content of these fungi.

The muscarinic agonist pilocarpine (15) was shown by Cushing (1932) to produce autonomic and somatomotor effects when injected intraventricularly into man or the rabbit. The same drug has since been shown by Monnier and Romanowski (1962) and by Riehl, Paul-David and Unna (1962) to produce an arousal response in animals when administered intravenously. This response is blocked by atropine. Zablocka and Esplin (1964) studied the actions of pilocarpine in cats and found a variety of effects both stimulant and depressant, all apparently mediated through cholinergic mechanisms.

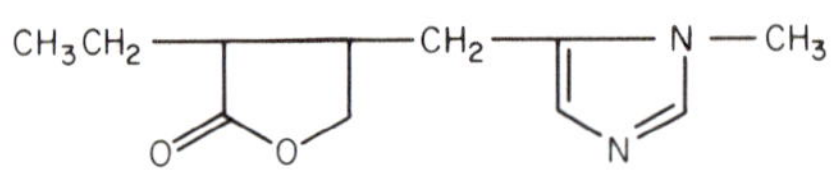

15

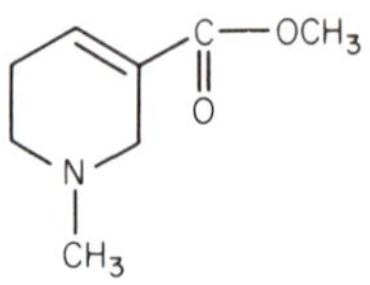

16

Arecoline (see formula 16), another muscarinic agonist from the betel-nut, the seed of *Areca catechu*, was first investigated in 1894 by Frohner who noted that it produced tremors in a number of species of animals.

Subsequently the central origin of this tremorogenic effect was confirmed by Holmstedt and Lundgren (1966).

Most potent muscarinic agents are quaternary ammonium salts and thus do not enter the central nervous system but the discovery of tremorine (see formula 17), by Everett, Blockus and Shepperd in 1956 gave an opening into a whole new field of tertiary amine muscarinic agents. Drugs of this type, particularly the active metabolite of tremorine, oxotremorine (see formula 18) (Cho, Haslett and Jenden, 1962) will be discussed in more detail later in this monograph.

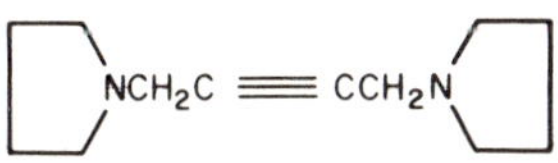

17

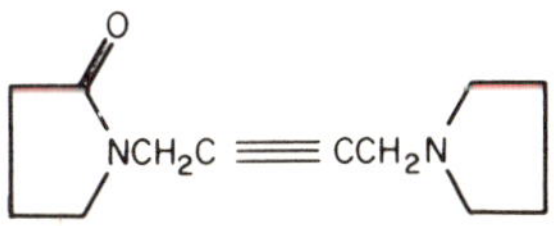

18

The behavioural effects produced by atropine, hyoscine and other belladonna alkaloids have been recognised since ancient times. Typical symptoms of belladonna poisoning include disturbances of memory and of orientation, hallucinations and often delirium; that is, the alkaloids produce a toxic psychosis. Such episodes have been reported following ingestion of the seeds and berries of *Atropa belladonna* and related plants but also after the instillation of atropine eyedrops, the use of belladonna plasters, and the ingestion of sleeping preparations containing hyoscine.

New interest was added to the central actions of anticholinergic drugs with the discovery in 1958 by Abood, Ostfeld and Biel of a series of piperidyl benzilates (see formula 19) with marked psychotomimetic activity. These agents will be discussed in considerably more detail later in this monograph. The main point of controversy since their discovery has been as to whether or not their central actions are due to competitive antagonism of acetylcholine at central cholinergic synapses.

Another central action of atropine and related drugs which will be discussed later is their action in reversing the central respiratory depression which results from poisoning with anticholinesterase agents.

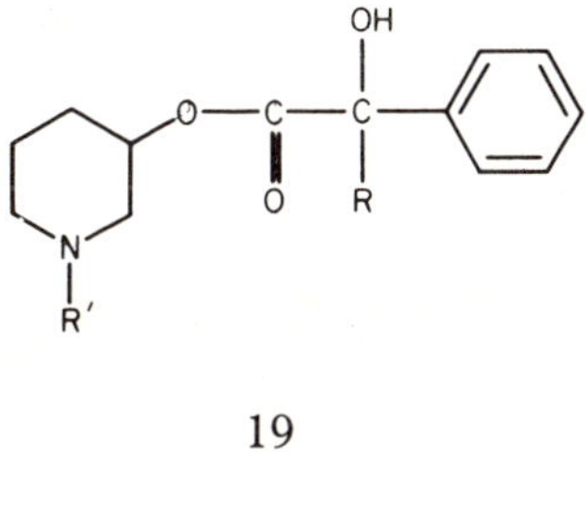

19

The central actions of nicotinic agonists, certainly of the parent compound, nicotine, must presumably have been recognised since this drug was first used in the form of tobacco. The alkaloid lobeline (see formula 20) is the active principle of a herb *Lobelia inflata* which was used in American medicine in the nineteenth century for its emetic, nauseant, expectorant and antiasthmatic actions (Volle and Koelle, 1970), most of which presumably originated in the central nervous system.

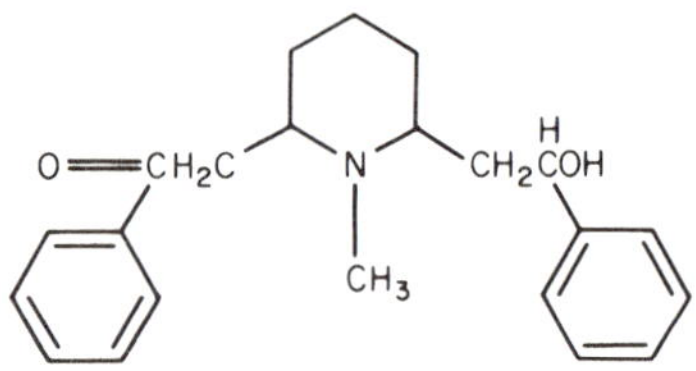

20

Very many studies have also been carried out subsequently with nicotine itself and some of these will be mentioned later in this monograph. Historically some of the earliest investigations were by van Praag (1855) who described the production of tremor by nicotine; by Nice (1913) who studied effects of the drug on spontaneous activity in mice; by Macht and Bloom (1920) who investigated effects of learning and by Behrend and Thienes (1933) who used conditioned response procedures.

Most nicotinic antagonists are quaternary ammonium salts and therefore devoid of central actions. In this context it is somewhat surprising that there has been, over the years, some discussion as to whether curare has central actions. That this is not the case appears to have been shown quite conclusively by Smith, Brown, Toman and Goodman (1947) who studied the effects on a volunteer of a dose of d-tubocurarine sufficient to induce complete muscular paralysis. No central functions appeared to be inspired in any way.

The tertiary and secondary amine ganglion blocking drugs of the mecamylamine or pempidine type might be expected to gain access to the central nervous system and there to produce effects. Indeed, it has been reported (Volle and Koelle, 1970), that tremors, mental confusion, seizures, mania or depression can result from the use of these drugs. Whether or not these actions result from blocking actions at central cholinergic synapses has not been resolved.

Central actions of anticholinesterase agents have been widely studied. The results of such studies are not always easy to interpret probably because the effects are often dose-dependent with stimulant actions at low doses being succeeded by depressant actions at higher doses.

Characteristically anticholinesterase agents produce an electrocortical arousal; that is, a change in the pattern of the electroencephalogram (EEG) or electrocorticogram (ECoG) to one of low amplitude, high frequency waves. This was shown with dyflos (see formula 21) by Wescoe, Green, McNamara and Krop (1948) and with physostigmine by Bremer and Chatonnet (1949). In each case atropine reversed the effect.

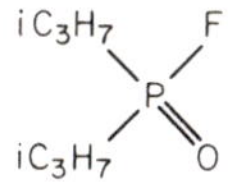

21

In 1953 Bradley and Elkes reported that the electrocortical arousal in cats given physostigmine was not accompanied by a behavioural arousal; that is, there was a dissociation between electrocortical activity and behaviour. The implication of these findings seems to be that a cholinergic system is involved in electrocortical arousal and that acetylcholine receptors of the muscarinic type are involved. This matter is discussed in greater detail later in the monograph.

Behavioural studies, which are also reviewed later, tend, on the whole, to suggest that behavioural changes only occur in animals in which anticholinesterase agents have produced a significant depletion in levels of brain acetylcholinesterase.

The effects of anticholinesterase agents on respiration demonstrate very clearly their dose-dependent dual action. Schweitzer and Wright (1938) investigated the effects of neostigmine and physostigmine on anaesthetised cats. Small doses potentiated respiration while higher doses had a depressant action, usually followed by a period of stimulation. These effects are generally assumed to be due to actions on the medullary respiratory centres and are reversed by atropine.

Review articles dealing with the central actions of anticholinesterases in some detail have been written by Holmstedt (1959), Machne and Unna (1963) and Karczmar (1970).

REFERENCES

Abood, L. G., Ostfeld, A. M. and Biel, J. H. (1958). A new group of psychotomimetic agents. *Proc. Soc. exp. Biol., N.Y.*, **97**, 483–486

Ambache, N. (1949). The peripheral action of Clostridium botulinum toxin. *J. Physiol., Lond.*, **108**, 127, 141

Ariens, E. J. (1954). Affinity and intrinsic activity in the theory of competitive inhibition. I. Problems and theory. *Arch. int. Pharmac.*, **99**, 32–49

Baeyer, A. (1867). Ueber das Neurin. *Justus Liebigs Annin. Chem.*, **142**, 322–326

Barlow, R. B. (1964). *Introduction to Chemical Pharmacology*, Methuen & Co. Ltd., London

Barlow, R. B. and Ing, H. R. (1948). Curare-like action of polymethylene bis-quaternary ammonium salts. *Br. J. Pharmac. Chemother.*, **3**, 298–304

Behrend, A. and Thienes, C. H. (1933). The development of tolerance to nicotine by rats. *J. Pharmac. exp. Ther.*, **48**, 317–325

Bernard, C. (1856). Analyse physiologique des propriétés des systèmes musculaires et nerveux au moyer du curare. *C.r. Lebh. Seanc. Acad. Sci., Paris*, **43**, 825–829

Bernard, C. (1857). *Leçons sur les effets des substances toxiques et médicamenteuses.* J.-B. Ballière et Fils, Paris

Bovet, D., Bovet-Nitti, F. and Marini-Bettolo, G. B. (1959). *Curare and Curare-like Agents.* Elsevier Publishing Co., Amsterdam

Bradley, P. B. (1968). Synaptic transmission in the central nervous system and its relevance for drug action. *Int. Rev. Neurobiol.*, **11**, 1–56

Bradley, P. B. and Elkes, J. (1953). The effects of some drugs on the electrical activity of the brain. *Brain*, **80**, 77–117

Bremer, F. and Chatonnet, J. (1949). Acetylcholine et cortex cérébrale. *Arch. int. Physiol.*, **57**, 106–109

Burn, J. H. and Dale, H. H. (1915). The action of certain quaternary ammonium bases. *J. Pharmac. exp. Ther.*, **6**, 417–438

Cho, A. K., Haslett, W. L. and Jenden, D. J. (1962). The peripheral actions of oxotremorine, a metabolite of tremorine. *J. Pharmac. exp. Ther.*, **138**, 249–257

Clark, A. J. (1933). *The Mode of Action of Drugs on Cells.* Edward Arnold, London

Clermont, P. de. (1854). Chimie organique—Note sur la préparation de quelques éthers. *C.r. Lebd. Seanc. Acad. Sci., Paris*, **39**, 338–347

Corne, S. J. and Edge, N. D. (1958). Pharmacological properties of pempidine (1:2:2:6:6-pentamethyl piperidine) a new ganglion blocking compound. *Br. J. Pharmac. Chemother.*, **13**, 339–349

Crum Brown, A. and Fraser, T. R. (1868). On the connection between chemical constitution and physiological action. Pt I. *Trans. R. Soc. Edinb.*, **25**, 151–203

Curtis, D. R. and Eccles, R. M. (1958). The excitation of Renshaw cells by pharmacological agents applied electrophoretically. *J. Physiol. Lond.*, **141**, 435–445

Cushing, H. (1932). *Papers relating to the pituitary body, hypothalamus and parasympathetic nervous system.* Charles C. Thomas, Springfield, Illinois

Dale, H. H. (1914). The actions of certain esters and ethers of choline, and their relation to muscarine. *J. Pharmac. exp. Ther.*, **6**, 147–190

Dickson, E. C. and Sheuky, R. (1923). Studies on the manner in which the toxin of Clostridium botulinum acts upon the body. I. The effect upon the autonomic system. *J. Exp. Med.*, **31**, 711–731

Dixon, W. E. (1907). On the mode of action of drugs. *Med. Mag. Lond.*, **16**, 454–457

Edmunds, C. W. and Long, P. H. (1923). Contribution to the pathologic physiology of botulism. *J. Amer. Med. Ass.*, **81**, 545–547

Engelhart, E. and Loewi, O. (1930). Fermentative Azetylcholinspaltung im Blut und ihre Hemmung durch Physostigmine. Naunym-Schmiedebergs. *Arch. exp. Path. Pharmak.*, **150**, 1–13

Everett, G. M., Blockus, L. E. and Sheppard, I. M. (1956). Tremor induced by tremorine and its antagonism by anti-parkinsonian drugs. *Science*, **124**, 79

Fabing, H. D. (1956). On going berserk: A neurochemical enquiry. *Scient. Mon. N.Y.*, **82**, 232–237

Feldberg, W. and Krayer, O. (1933). Das Auftreten eines azetylcholinartigen Stoffes in Herzvenenblut von Warmblütern Bei Reizung der Nervi vagi. *Arch. exp. Path. Pharmak.*, **172**, 170–193

Frohner, E. (1894). Untersuchungen über das Arecolin, ein Ersatzmittel für Pilocarpin und Eserin. *Monatsch. Theirheilkunde*, **5**, 353–362

Holmstedt, B. (1959). Pharmacology of organophosphorus cholinesterase inhibitors. *Pharmac. Rev.*, **11**, 567–688

Holmstedt, B. (1963). Structure activity relationships of the organophosphorus anticholinesterase agents. *Cholinesterases and Anticholinesterase Agents. Handb. exp. Pharmak.*, Supple. 15. (ed. G. B. Koelle), Springer-Verlag, Berlin, pp 428–485

Holmstedt, B. and Lundgren, G. (1966). Tremorogenic agents and brain acetylcholine. *Mechanism of Release of Biogenic Amines*. (eds. U. S. von Euler, S. Rossel and B. Uvnas) Pergamon, London, pp 439–468

Hunt, R. and Taveau, R. de M. (1906). On the physiological action of certain choline derivatives and new methods for detecting choline. *Br. Med. J.*, **2**, 1788–1791

Innes, I. R. and Nickerson, M. (1965). *The Pharmacological Basis of Therapeutics*. (eds. L. S. Goodman and A. Gilman) The Macmillan Co., New York, pp 521–545

Jobst, J. and Hesse, O. (1864). Über die Bohne von Calabar. *Ann. der. Chem. und. Pharm.*, **129**, 115–121

Karczmar, A. G. (1970). *Anticholinesterase Agents; International Encyclopedia of Pharmacology and Therapeutics*, Sect. 13, Vol. I. (ed. A. G. Karczmar), Pergamon Press, Oxford

King, H. (1935). Curare Alkaloids. I. Tubocurarine. *J. chem. Soc.*, 1381–1389

Kleinwachter, L. (1864). Beobachtung über die Wirkung des Calabar—Extracts gegen Atropin-Vergiftung. *Berl. Klin. Wschr.*, **1**, 369–371

Koelle, G. B. (1970a). Neurohumoral transmission and the autonomic nervous system. *The Pharmacological Basis of Therapeutics.* 4th edition (eds. L. S. Goodman and A. Gilman) The Macmillan Co., New York, pp 402–441

Koelle, G. B. (1970b). Anticholinesterase agents. *The Pharmacological Basis of Therapeutics.* 4th edition (eds. L. S. Goodman and A. Gilman) The Macmillan Co., New York, pp 442–465

Koelle, G. B. (1970c). Neuromuscular blocking agents. *The Pharmacological Basis of Therapeutics.* 4th edition (eds. L. S. Goodman and A. Gilman) The Macmillan Co., New York, pp 601–619

Koelle, G. B. (1970d). Parasympathomimetic agents. *The Pharmacological Basis of Therapeutics.* 4th edition (eds. L. S. Goodman and A. Gilman) The Macmillan Co., New York, pp 466–477

Lange, W. and Krueger, G. von (1932). Über Ester der Monofluorphosphorsäure. *Ber. dt. Chem. Ges.*, **65**, 1598–1601

Langley, J. N. (1878). On the physiology of the salivary secretion. Pt II. On the mutual antagonism of atropine and pilocarpine having especial reference to their relations in the submaxillary gland of the cat. *J. Physiol. Lond.*, **1**, 339–376

Langley, J. N. and Dickinson, W. L. (1889). On the local paralysis of peripheral ganglia and on the connection of different classes of nerve fibres with them. *Proc. Roy. Soc.*, **46**, 423–431

Loewi, O. (1921). Über humorale Übertragbarkeit der Herznervenwirkung. *Pflugers Arch. ges. Physiol.*, **189**, 239–242

Loewi, O. and Navratil, E. (1926). Über humorale Übertragbarkeit der Herznervenwirkung. X. Mitteilung Uber das Schicksal des Vagusstoffs. *Pflugers. Arch. ges. Physiol.*, **214**, 678–688

Machne, X. and Unna, K. R. W. (1963). Actions at the central nervous system. *Cholinesterases and Anticholinesterase Agents. Handb. exp. Pharmak.*, Supple. 15. (ed. G. B. Koelle), Springer-Verlag, Berlin, pp 679–700

Macht, D. I. and Bloom, W. (1920). Comparative study of ethanol, caffeine and nicotine on behaviour of albino rats. *Proc. Soc. exp. Biol. N.Y.*, **18**, 99–100

MacIntosh, F. C., Birks, R. I. and Sastry, P. B. (1956). Pharmacological inhibition of acetylcholine synthesis. *Nature, Lond.*, **178**, 1181

Matthes, K. (1930). The action of blood on acetylcholine. *J. Physiol. Lond.*, **70**, 338–348

Monnier, M. and Romanowski, W. (1962). Les systèmes cholinoceptif cérébraux—actions de l'acetylcholine, de la physostigmine, pilocarpine et de gaba. *Electroenceph. clin. Neurophysiol.,* **14**, 486–500

Nice, L. B. (1913). Studies on the effects of alcohol, nicotine and caffeine on white mice. II. Effects on activity. *J. exp. Zool.,* **14**, 123–151

Paton, W. D. M. (1961). A theory of drug action based on the rate of drug-receptor combination. *Proc. Roy. Soc. B.,* **154**, 21–69

Paton, W. D. M. and Zaimis, E. J. (1949). The pharmacological actions of polymethylene bistrimethylammonium salts. *Br. J. Pharmac. Chemother.,* **4**, 381–400

Paton, W. D. M. and Zaimis, E. J. (1951). Paralysis of autonomic ganglia by methonium salts. *Br. J. Pharmac. Chemother.,* **6**, 155–168

Riehl, J. L., Paul-David, J. and Unna, K. R. (1962). Comparison of the effects of arecoline and muscarine on the autonomic nervous system. *Int. J. Neuropharmac.,* **1**, 393–401

Rylant, P. (1927). La transmission humorale de l'action des nerfs cardiaques de Loewi chez la mamière. *C.r. Seanc. Soc. Biol.,* **96**, 204–205

Schmiedeberg, O. and Koppe, R. (1869). *Das Muscarin, das giftige Alkaloid des Fliegenpilzes.* Vogel, Leipzig

Schrader, G. (1952). *Die Entwicklung neuer Insektizide auf Grundlage von organischen Fluor- und Phosphorverbindungen.* Monographie No. 62. Aufl. Weinheim, Verlag Chemie.

Schueler, F. W. (1955). A new group of respiratory paralyzants. *J. Pharmac. exp. Ther.,* **115**, 127–143

Schweitzer, A. and Wright, S. (1938). Action of prostigmine and acetylcholine on respiration. *Quart. J. exp. Physiol.,* **28**, 33–47

Smith, S. M., Brown, H. O., Toman, J. E. P., and Goodman, L. S. (1947). The lack of cerebral effects of d-tubocurarine. *Anesthesiology,* **8**, 1–14

Spinks, A., Young, E H. P., Farrington, J. A. and Dunlop, D. (1958). The pharmacological actions of pempidine and its ethyl homologue. *Br. J. Pharmac. Chemother.,* **13**, 501–520

Stedman, E. and Barger, G. (1925). Physostigmine (eserine) Pt. III. *J. Chem. Soc.,* **127**, 247–258

Stephenson, R. P. (1956). Modification of receptor theory. *Br. J. Pharmac. Chemother.,* **11**, 379–393

Stone, C. A., Torchiana, M. L., Navarro, A. and Beyer, K. H. (1956). Ganglionic blocking properties of 3-methylamino-isocamphane hydrochloride (mecamylamine): A secondary amine. *J. Pharmac. exp. Ther.,* **117**, 169–183

Van Praag, J. L. (1855). IV. Nicotin. *Toxikologisch—pharmakodynamische Studein Virschows Arch.*, **8**, 56–102

Volle, R. L. and Koelle, G. B. (1970). Ganglionic stimulating and blocking agents. *The Pharmacological Basis of Therapeutics.* 4th edition (eds. L. S. Goodman and A. Gilman) The Macmillan Co., New York, pp 585–600

Wescoe, W. C., Green, R. E., McNamara, B. P. and Krop, S. (1948). The influence of atropine and scopolamine on the central effects of DFP. *J. Pharmac. exp. Ther.*, **92**, 63–72

Zablocka, B. and Esplin, D. W. (1964). Central excitatory and neuromuscular paralysant effects of pilocarpine in cats. *Arch. int. Pharmacodyn.*, **147**, 490–496

2 Drug Actions at Peripheral Muscarinic Sites

2.1 Agonists

The work of Dale in 1914 established that the alkaloid muscarine mimicked the actions of acetylcholine at postganglionic parasympathetic endings. Agonist drugs which have this action are said to be muscarinic. The most common method of assay for such drugs involves the use of a piece of isolated gut (usually guinea pig ileum or rabbit duodenum) suspended in a suitable oxygenated medium maintained at body temperature. Agonists cause contractions of the tissue which are prevented or abolished by atropine or other antimuscarinic drugs. There are also a number of available *in vivo* methods of assay. One of the more commonly used makes use of the action of these drugs in causing vasodilatation and hence sharp falls in blood pressure of anaesthetised or spinal animals, usually cats or rats. Another commonly used method involves the measurement in rats of saliva production which is stimulated by muscarinic drugs. Whatever assay method is used a check must always be made that the response is prevented or abolished by atropine.

In vitro techniques, using isolated tissues, are of greatest value in giving estimates of the activity of drugs relatively uncomplicated by such factors as rate of absorption, distribution, metabolism or excretion. *In vivo* techniques, on the other hand, give better indications of the activities of drugs which may be intended for clinical use although such use of these drugs is not widespread. Methacholine (acetyl-β-methylcholine) (see formula 22) has been used to some extent to arrest attacks of paroxysmal tachycardia and to treat glaucoma. Bethanechol (β-methylcholine carbamate) (see formula 23) and carbachol (carbamylcholine) (see formula 24) have been used in the treatment of urinary atony and distension, and intestinal atony. Carbachol has also been used for glaucoma.

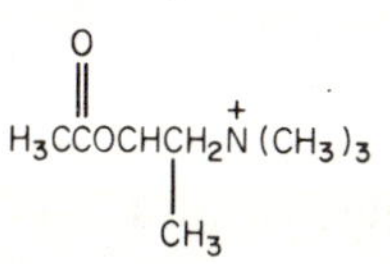

22

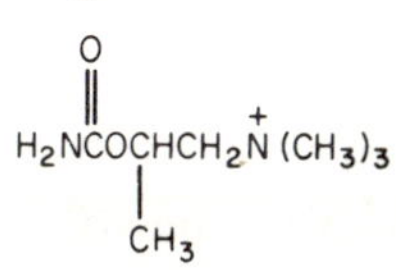

23

$$H_2N\overset{\overset{O}{\|}}{C}OCH_2CH_2\overset{+}{N}(CH_3)_3$$

24

There has been considerable discussion concerning the structural requirements for agonist activity at muscarinic sites and several groups of workers have proposed models for the muscarinic receptor based on structure-activity studies with agonist drugs. This kind of approach assumes that there are complementary groups on the receptor capable of interaction with those groups on the drug which prove to be essential for high activity.

No attempt will be made here to review the whole field of peripherally-acting muscarinic agonists. Adequate reviews have been written by Waser (1961); Barlow (1964); Bebbington and Brimblecombe (1965); Martin-Smith, Smail and Stenlake (1967); and Kier (1970a). In this chapter attention will be focused on these pieces of work which are of greatest relevance to the discussion in Chapter 5 of drug actions which may be mediated through muscarinic receptors in the central nervous system.

Clearly any attempt to delineate the muscarinic receptor must take account of the structure of the natural transmitter, acetylcholine (see formula 25) and it soon became apparent from early studies using various derivatives of acetylcholine that the quaternary ammonium group seemed to be essential for activity at this site. However the nature of the groups attached to the quaternised nitrogen atom were also clearly of major significance and an important general rule, the so-called "five-atom rule" was propounded by Ing in 1949. This stated that for maximal activity there should be attached to the nitrogen atom, as well as three methyl groups, a fourth group with a chain of five atoms (for example the C-C-O-C-C chain in acetylcholine).

$$CH_3\overset{\overset{O}{\|}}{C}OCH_2CH_2\overset{+}{N}(CH_3)_3$$

25

The necessity for three methyl groups to be attached to the nitrogen atom for high activity had been shown earlier by Stehle, Melville and Oldham (1936) who successively replaced methyl groups in the acetylcholine molecule by 1, 2 and 3 hydrogen atoms. The compounds so produced were studied for muscarinic activity using as measures of activity the hypotensive response in anaesthetized cats, the contraction of isolated rabbit intestine and the reduction in rate of beat of the isolated frog's heart. Replacement of one methyl group by a hydrogen atom resulted in a forty to fifty-fold drop in activity; replacement of two methyl groups gave a five hundred to one-thousand-fold drop and replacement of all three groups gave a drop in activity of two-thousand to forty-thousand times depending on the test used.

The five-atom rule received convincing confirmation from the work of, among others, Stephenson (1956) and van Rossum and Ariens (1959) who measured the muscarinic activities of homologous trimethylammonium salts (see formula 26) and found maximum activity with the amyl compound (formula 26, n = 4).

$$CH_3(CH_2)_n\overset{+}{N}(CH_3)_3$$

26

Their results are summarised in Table 2.1.

TABLE 2.1

The muscarinic activities of alkyl trimethylammonium salts $CH_3(CH_2)_n\overset{+}{N}(CH_3)_3$

n	Potency as percentage of that of the amyl compound	
	Guinea-pig ileum Stephenson (1956)	Rat jejunum van Rossum and Ariens (1959)
0	2.1	3.2
1	1.5	2.0
2	1.1	1.6
3	47	63
4	100	100
5	—	40

Another interesting series of compounds was examined by Ing, Kordik and Tudor-Williams (1952). In these they replaced a methylene group in the polymethylene chain of alkyl trimethylammonium salts with either an ether oxygen or a carbonyl group. Compounds with an ether oxygen were always much more potent than those with a carbonyl group. Maximal activity was found in the compound with the ether oxygen in position 3 of the chain (see formula 27).

$$CH_3CH_2OCH_2CH_2\overset{+}{N}(CH_3)_3$$

27

Activity was increased when the ether oxygen atom was incorporated into a furan ring (see formula 28). The compound shown had an activity twenty to forty times greater than that of the pentyltrimethylammonium salt. There is a clear structural resemblance between this compound and that of muscarine itself (see formula 4) and of the 1,3-dioxolan (see formula 29) described by Fourneau, Bovet, Bovet and Montezin (1944). The latter is a very active muscarinic agent with an activity about ten times that of acetylcholine when measured on the isolated guinea pig ileum. In comparison (+)-muscarine was reported by Gyermak and Unna (1958) to be three times more potent than acetylcholine.

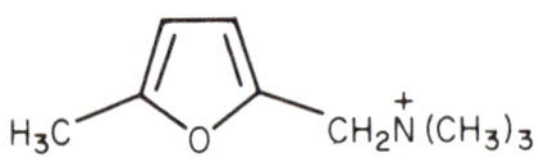

28

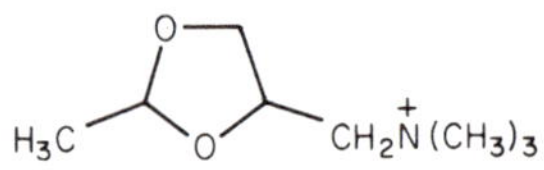

29

It is worth noting that in cyclic ethers like compounds 27 and 29 the presence of the 2-methyl group is necessary for high activity. This is added evidence in favour of the five-atom rule. It has been suggested that the presence of this group increases the electron density on the ether oxygen atom (Beckett, Harper, Clitherow and Lesser, 1961) or causes an increase in van der Waals binding forces (Waser, 1961).

From the work outlined above and from many related studies it was apparent that a trimethylammonium group and an ether oxygen atom were present in most active muscarinic agonists. Waser in 1961, as a result of studies of muscarine (see formula 4), muscarone (see formula 30) and related compounds, concluded that maximal activity at the muscarinic receptor depended mostly on the cationic nitrogen and a nucleophilic group (for example, the ether oxygen of muscarine or the carbonyl oxygen of muscarone) separated by about 4 Å.

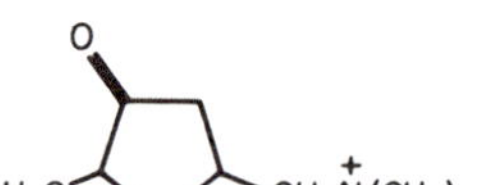

30

A rather different picture emerged from the work of Beckett and his colleagues (Beckett *et al.*, 1961; Beckett, Harper and Clitherow, 1963). They considered the stereochemistry and molecular dimensions of muscarine and acetyl α- and β-methyl-cholines (see formulae 31 and 32), and concluded that the main groups involved in drug receptor interaction were the quaternary ammonium group and the ether oxygen atom, separated by about 3 Å. These would interact with postulated sites 1 and 2

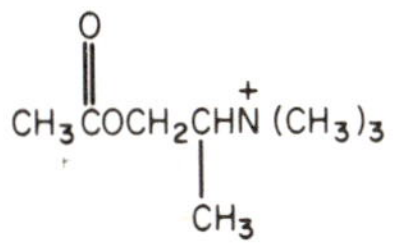

31

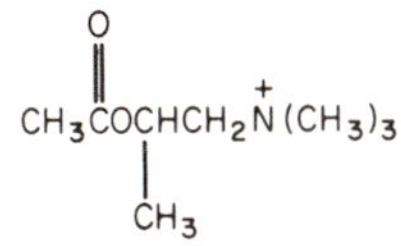

32

on the receptor. The hydroxyl group of muscarine which is situated some 5–7 Å from the quaternary ammonium group was considered to interact with a subsidiary site 3 on the receptor. This third site would also, presumably, be concerned in interaction with the carbonyl group of acetylcholine, the ether oxygen atom of the dioxolan (see formula 29) or the double bond of the furan (see formula 28). Measurements of molecular models reveal that to accommodate any of these groups the receptor site 3 would need to be large. The muscarinic receptor, as envisaged by Beckett and his colleagues, is shown in Figure 2.1.

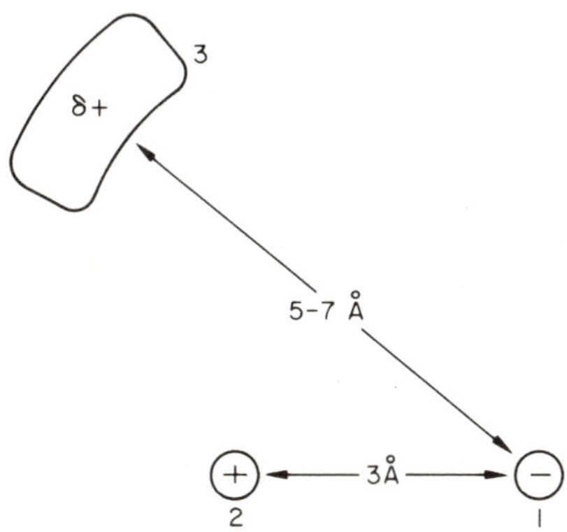

FIGURE 2.1 *The muscarinic receptor* (after Beckett, Harper and Clitherow, 1963)

A basic difference between the ideas of Waser and of Beckett and his co-workers concerns muscarone (see formula 30). As was stated above, Waser considered that the carbonyl oxygen of muscarone interacted with the same receptor site as the ether oxygen of muscarinic. Beckett *et al.*, on the other hand, suggested that their postulated receptor site 2 accommodated the ether oxygen atom of both muscarine and muscarone. This extra binding between drug and receptor might thus account for the fact that (±) – muscarone is about three times more active than (±) – muscarine. Beckett *et al.* also stressed the apparent planar nature of the receptor and noted that the greater planarity of muscarone might also facilitate its binding to the receptor.

There is an interesting difference in the potencies of the enantiomers of muscarine and muscarone. (+)-Muscarine is between two hundred and eight hundred times more potent than (−)-muscarine (Hardeggar and Lohse, 1957) whereas the more active (−)-2R-5R-isomer of muscarone is only three times more active than its enantiomorph. This may be due to the fact that the (−) form of muscarone, unlike the (+) form of muscarine would not fit the receptor as proposed by Beckett, Harper and Clitherow. Waser (1961) suggested an alternative explanation whereby (−)-muscarone would adopt a folded form at the receptor thus enabling its carbonyl group to interact with the second receptor site. Belleau and Puranen (1963) pointed out, however, that such a folded form of (−) muscarone would represent an unfavourable conformation.

The discovery of oxotremorine (see formula 18), the active metabolite of tremorine (see formula 17) provided a new and valuable tool for studies into the nature of the muscarinic receptor. In their original paper describing the peripheral pharmacology of oxotremorine Cho, Haslett and Jenden (1962) noted that this drug was approximately equiactive with acetylcholine as a muscarine agonist. Its structure, however, was very different from that of other muscarinic agonists and Cho, Haslett and Jenden suggested that its activity might be explained on the basis of an "induced fit" hypothesis similar to that proposed for the interaction of enzymes with their substrates (Koshland, 1959).

However, an alternative explanation for the activity of oxotremorine was proposed by Bebbington, Brimblecombe and Shakeshaft (1966). They noted that, because of free rotation about the C≡C−CH_2 bond, the compound can exist in two possible planar forms, cisoid (see formula 33) or transoid (see formula 34).

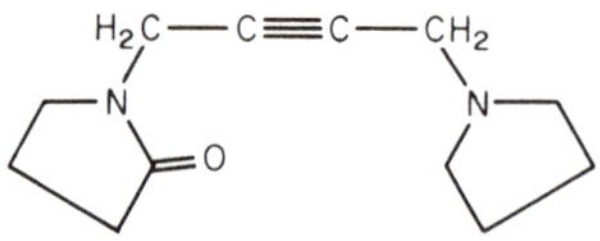

33

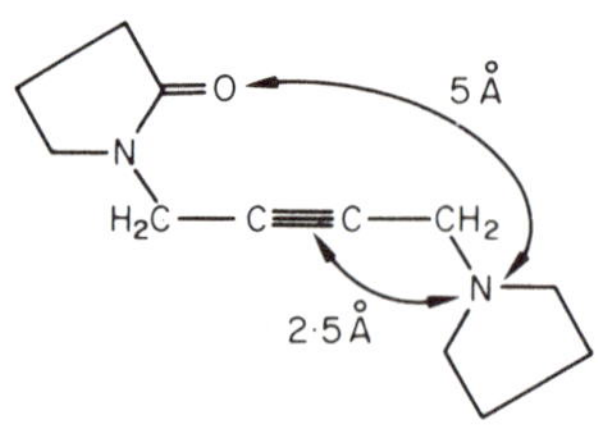

34

Measurement of molecular models revealed that in the transoid form the distance between the pyrrolidine nitrogen atom and the carbonyl oxygen atom was about 5 Å, and the distance between the pyrrolidine nitrogen and the acetylenic bond was about 2.5 Å. Thus if it were assumed that the groups involved in the interaction between drug and receptor were the pyrrolidine nitrogen, the acetylenic bond and the carbonyl group, these,

TABLE 2.2

The muscarinic activities of oxotremorine and related compounds

Compound	Muscarinic activity Acetylcholine = 1 (Isolated Guinea pig ileum)
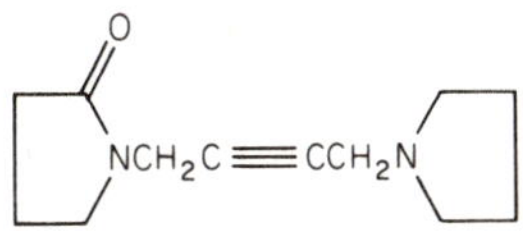 (Oxotremorine)	1.48
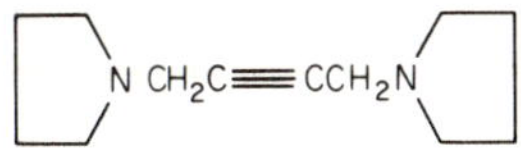 (Tremorine)	0.01
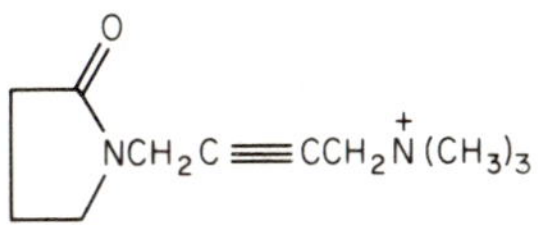	1.74
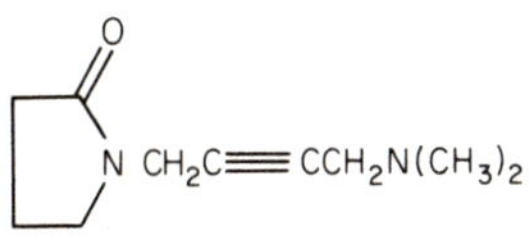	0.16
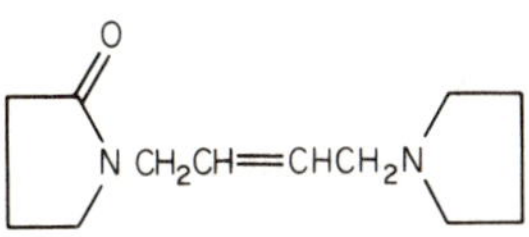	0.01
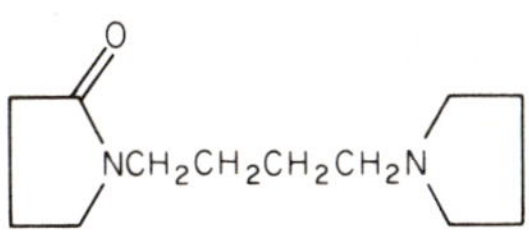	0·01
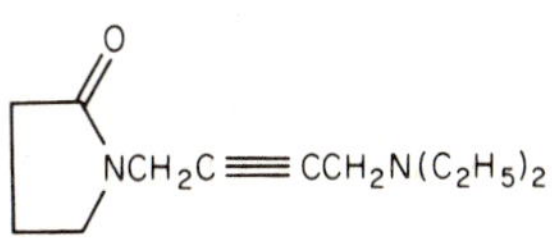	0·01
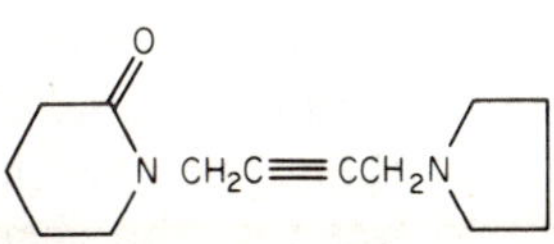	Antagonist

with oxotremorine in its transoid form, would be well placed to interact respectively with Sites 1, 2 and 3 of the receptor proposed by Beckett, Harper and Clitherow (1963); Bebbington, Brimblecombe and Shakeshaft (1966) and Bebbington and Brimblecombe (1965) suggested that the acetylenic bond represents a region of high electron density capable of interaction with the same receptor site as the ether oxygen atom of acetylcholine or muscarine.

A number of compounds related to oxotremorine were synthesised and tested for muscarinic activity. Some results are given in Table 2.2.

From the results listed in Table 2.2 several points emerge:

(1) The carbonyl group in oxotremorine is essential for activity as evidenced by the much higher potency of oxotremorine than tremorine.

(2) The pyrrolidine nitrogen forms a satisfactory replacement for a quaternary ammonium group resulting in only a very small loss of activity (from 1.74 to 1.48).

(3) The acetylenic bond is essential for activity in compounds of this type. Partial or complete saturation of the bond results in complete loss of activity.

(4) Increase in size of the lactam ring results in a change from agonist to antagonist activity.

(5) Replacement of the pyrrolidine ring with a dimethylamino group results in partial loss in activity. Activity is almost completely lost when the N-alkyl substituents are higher than methyl.

Bebbington, Brimblecombe and Shakeshaft (1966) then examined the muscarinic activities of other acetylenic amines. In particular the compound which had been reported by Jacob, Marszak, Bardsia, Marszak-Fleury and Epsztein (1952) was resynthesised and studied (see formula 35).

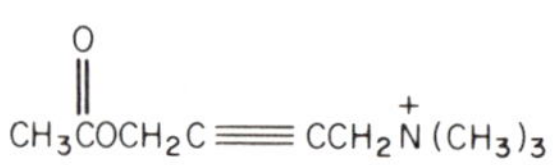

35

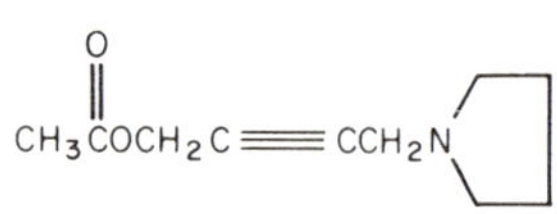

36

Its potency, as measured on the isolated guinea-pig ileum was about one-half that of acetylcholine. Measurement of a molecular model revealed that its active groups (quaternary nitrogen atom, acetylenic linkage and carbonyl group) were spaced very similarly to the active groups in oxotremorine. Accordingly the pyrrolidine analogue (see formula 36) was made but its activity was very low being only about one two-hundredth that of acetylcholine. The obvious deduction was that the character of the carbonyl group was different in this compound and in oxotremorine. Accordingly a group of compounds was synthesised, with their active groups so arranged as to correspond to the distances proposed in the receptor model of Beckett, Harper and Clitherow (1963) but having carbonyl groups in different chemical environments. The results are shown in Table 2.3.

TABLE 2.3

Muscarinic activities of acetylenic amines R-$CH_2C{\equiv}CCH_2R'$ *and* C=O *absorption maxima of pyrrolidine tertiary amines*

R	Muscarinic activity Acetylcholine = 1 (Isolated guinea pig ileum) $R' = \overset{+}{N}(CH_3)_3$,	$R' = N$	C = O absorption maximum cm^{-1}
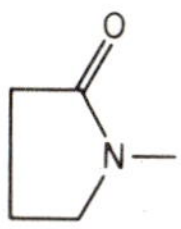	1.74	1.48	1689
(±) 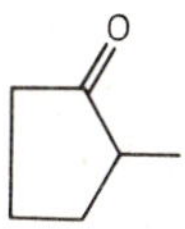	0.11	0.003	1733
(±) 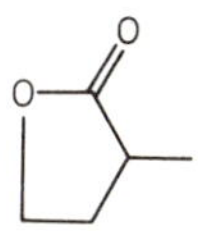	0.06	0.001	1773
$CH_3CON(CH_3)-$	0.87	1.15	1647
$(CH_3)_2NCON(CH_3)-$	0.03	0.04	1652
$(CH_3)_2NCOO-$	0.15	0.004	1700
CH_3COCH_2-	2.10	0.005	1715
CH_3COO-	0.06	0.001	1736

Variations are apparent in the activities of the quaternary ammonium salts listed in Table 2.3 but the variation is small in comparison with that shown by the tertiary bases. Probably steric and stereochemical factors contribute mainly to the differences in activity of the quaternary salts but the greater differences among the pyrrolidine tertiary amines seem to be connected with the character of the carbonyl group, for reference to Table 2.3 shows that as the frequency of the carbonyl absorption in the infrared region decreases so the activity of the tertiary base becomes closer to that of the corresponding quaternary salt. This decrease in frequency of absorption corresponds to a greater delocalisation of electrons in the C = O group; that is, there is an increase in the contribution of structure 38 to the total resonance of the molecule 37.

37 38

It was concluded from this study that the amine group and the carbonyl group contribute to the affinity of these agonists for the receptor. If the amine group is quaternised then it makes the major contribution to the affinity with the contribution of the carbonyl group being negligible. With tertiary bases, on the other hand, the contribution of the amine group to the total affinity of the molecule is relatively small and the contribution of the carbonyl group becomes of significance. Thus in ureas and amides, where the carbonyl absorption frequencies are relatively low the tertiary amines are almost as active as the quaternary ammonium salts.

From this work a structure for the muscarinic receptor was proposed (see figure 2.2), rather similar to that proposed earlier by Beckett and his colleagues but differing somewhat in dimensions and putting a different emphasis on the importance of site 3.

From studies of this kind it is only possible to conjecture concerning the nature of the receptor groups which are involved in interactions with the active groups on the drugs, but it seems almost certain that site 1 is an anionic site; site 2 has to be capable of interaction with an ether oxygen atom and an acetylenic bond and it has been suggested that a metal ion could be involved; the group at site 3 could well be capable of hydrogen bonding with the agonist.

More recently other groups of workers have made further studies of oxotremorine. For example, Martin-Smith, Smail and Stenlake (1967), reinterpreted the results obtained by Bebbington, Brimblecombe and Shakeshaft (1966) and suggested that the oxotremorine molecule existed

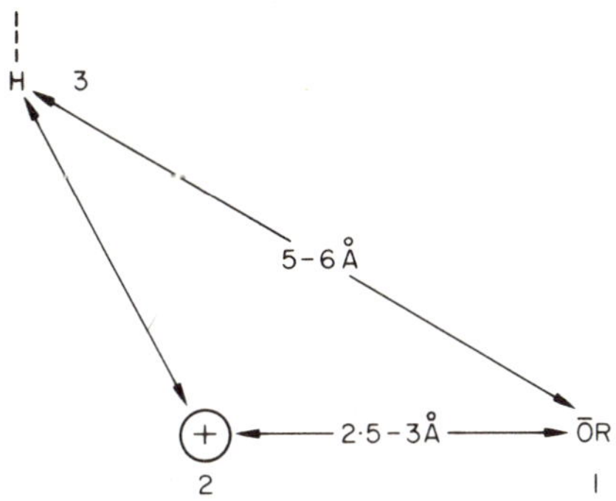

FIGURE 2.2 *The muscarinic receptor* (after Bebbington and Brimblecombe, 1965)

substantially in the form of formula 39 assuming that at physiological pH there is protonation of the pyrrolidine nitrogen atom.

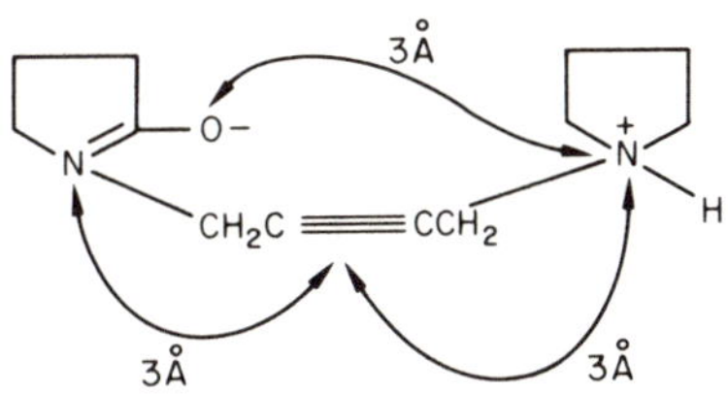

39

With this structure there would be the equivalent of two highly positively charged nitrogen atoms at approximately 3 Å from two centres of high electron density. This would then correspond well with the N – O acetoxy distance in the so-called quasi-ring form of acetylcholine suggested by Canepa (1965) (see formula 40).

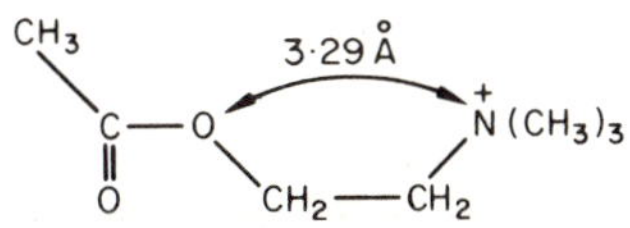

40

A more recent paper is by Kier (1970b) who calculated the conformational preference of oxotremorine using a method known as extended Huckel orbital theory. He showed that the oxotremorine molecule can exist in a variety of preferred structures, a point in agreement with that made by Martin-Smith, Smail and Stenlake (1967) who pointed out that there was no good reason to chose between either of the planar transoid forms or the two cisoid forms of the compound. Kier found that the two rings of oxotremorine exhibit considerable independence with respect to their conformational preference but the energy-permitted stereochemistry is such as to allow the quaternary nitrogen, the acetylenic bond and the carbonyl oxygen to assume the same relationship as had previously been suggested by Bebbington and Brimblecombe (1965) for what Kier describes as the "muscarinic pharmacophore". Kier was able to calculate the dimensions of a muscarinic pharmacophore based on conformations calculated for oxotremorine and those he had calculated previously (Kier, 1967) for acetylcholine, muscarine and muscarone. This pharmacophore is shown in figure 2.3 and it is possible to obtain from it the best idea of the receptor based on knowledge available at the present time.

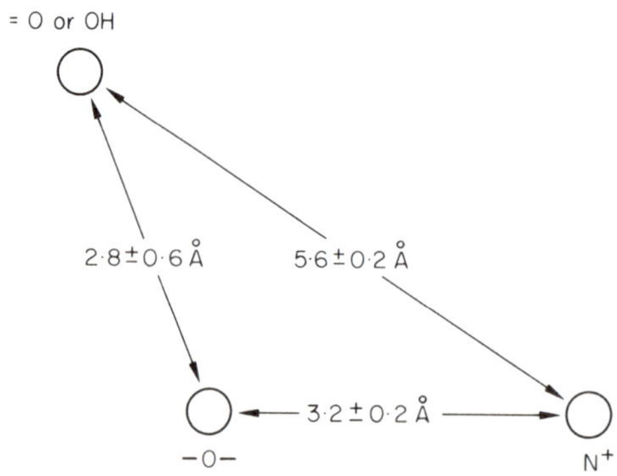

FIGURE 2.3 *The muscarinic pharmacophore* (after Kier, 1970b)

Later in this monograph evidence will be presented to suggest the presence in the central nervous system of muscarinic receptors, analogous to those in the periphery, with which tertiary amine muscarinic agonists like oxotremorine interact to produce certain effects, notably tremor. The foregoing discussion, if this hypothesis is correct, can therefore be taken to apply equally to these postulated central muscarinic receptors.

2.2 Antagonists

Antimuscarinic drugs, anticholinergic or antiacetylcholine drugs as they are usually now called act as competitive antagonists to the actions of acetylcholine and other agonists at postganglionic parasympathetic endings. There are both *in vivo* and *in vitro* methods for their assay. *In vivo* methods depend on antagonising some effect of exogenously administered agonists or on measuring an effect produced by the drug blocking the actions of endogenous acetylcholine. An example of the former type of test procedure is the blockade of oxotremorine-induced salivation, usually in mice. This gives an estimate of the peripheral antiacetylcholine potency of the antagonist and, in the same animal, an estimate of its central antiacetylcholine potency, which can be obtained by measuring the ability of the drug to block oxotremorine-induced tremors. This latter point will be discussed more fully later in the monograph. A commonly used method which does not involve exogenous administration of an agonist is to use the activity of the antiacetylcholine drug in producing mydriasis, also usually in mice, as a measure of its effectiveness. The mydriatic effect of antiacetylcholine drugs depends on their ability to block the postganglionic endings of the ciliary nerve which innervates the circular muscle of the iris (*constrictor pupillae*). Normally, tonic impulses along the ciliary nerve tend to keep these muscles more or less contracted but when an antiacetylcholine drug effects a blockade the pupil becomes dilated.

The most commonly used *in vitro* method for assaying antiacetylcholine activity makes use of the ability of the drugs to block contractions of the isolated guinea pig ileum induced by acetylcholine or other muscarinic agonist. The most accurate method is that described by Barlow, Scott and Stephenson (1963) in which regular contractions of the ileum are obtained by two different concentrations of acetylcholine and then the doses of acetylcholine are increased to give comparable contractions in the presence of a particular concentration of antagonist. It is thus possible to determine the dose ratio corresponding to this concentration of antagonist. (Dose ratio is equal to the dose of agonist required to produce a given response in the presence of an antagonist, divided by the dose required to produce the same response in the absence of the antagonist = A/a). The affinity constant of the antagonist can then be calculated from the equation $BK = A/a - 1$ (Gaddum, 1957) where B is the concentration of the antagonist and K its affinity constant. If plots of dose ratio against concentration are linear this indicates that the antagonist is acting competitively.

Earlier than this Schild (1947) had introduced the use of a measure which he called pAx to express antagonist activity and the value pA2 (the logarithm of the reciprocal of the concentration of antagonist which necessitates doubling the concentration of the agonist to obtain a constant response) has been widely used. pA2 is equal to the logarithm of K, the affinity constant, but is, under some circumstances, not such a precise measure as the latter especially when equilibrium conditions take relatively long times to become established. For a more detailed consideration of these matters and for a description of other methods for estimating antiacetylcholine activity reference should be made to Barlow (1964).

Clinically, antiacetylcholine drugs have a variety of uses. They are used to dilate the pupils to enable opthalmological examinations to be carried out and also in the treatment of certain eye conditions like acute iritis and keratitis. They reduce secretions into the respiratory tract in choryza and hayfever and they have some bronchial dilator action which is useful in bronchial asthma. As premedicants prior to anaesthesia they inhibit salivation and excess secretion into the airway and also produce a vagal block to prevent cardiac arrest due to increased vagal activity during surgery. In the treatment of peptic ulcer they reduce the movement of the gut and also the volume and the acidity of gastric secretion. The drugs have a general use in relieving the pain in various conditions like colic where there is spasmodic contraction of smooth muscle. Finally, their central actions are used to prevent motion sickness and in the treatment of Parkinson's disease. The latter use is considered in more detail later in this monograph.

Again no attempt will be made to review the whole field of peripheral muscarinic antagonists. Reference should be made to Kuznetsov and Golikov (1962) and to Barlow (1964). Here only some of the salient features, and some of particular interest to the author, will be considered.

Atropine (see formula 9) is the prototype of all antiacetylcholine drugs and, not surprisingly, much of the early work in this field was concerned with the study of compounds structurally closely related to atropine. Atropine is the racemic form of tropyltropine or hyoscyamine and it soon became apparent that stereochemical factors were important in the activity of compounds of this type. (—)-Hyoscyamine has been known to be more active than the (+)-isomer but the precise stereochemistry was not elucidated until 1961 by Fodor and Csepreghy who found that (−)-tropic acid has the S-configuration. The more active (−)-isomer of hyoscine (tropylscopine) (see formula 10) is also derived from S-tropic acid. Other, more general papers on this topic are by Ing (1955), Gyermek and Nador (1957) and Ariens (1966).

Most potent antiacetylcholine drugs can be considered to be derived from muscarinic agonists by the introduction of bulky substituent groups into the molecule. These groups are capable of van der Waals binding and so increase the affinity of the drug for the receptor. It has been suggested (Bebbington and Brimblecombe, 1965) that there is a relatively large area lying outside the area of the proposed three-site receptor where such van der Waals interaction can take place.

Examples of this relationship between the structures of agonists and antagonists are found with acetylcholine (see formula 25) and atropine (see formula 9); more obviously with acetylcholine and the benzilic acid derivatives (see formula 41) and related compounds (Abramson, Barlow, Mustafa and Stephenson, 1969); with the active muscarinic 1,3-dioxolan (see formula 29) (Fourneau, Bovet, Bovet and Montezin, 1944) and the 2,2-diphenyl-1,3-dioxolan derivative (see formula 42) (Brown and Werner, 1949) and with acetylenic amines such as formula 43 (Dahlbom, Karlen, Ramsby, Kraft and Mollberg, 1964) which are based on the oxotremorine structure.

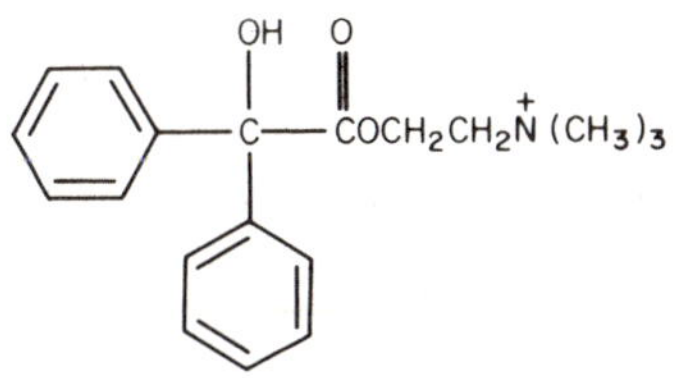

41

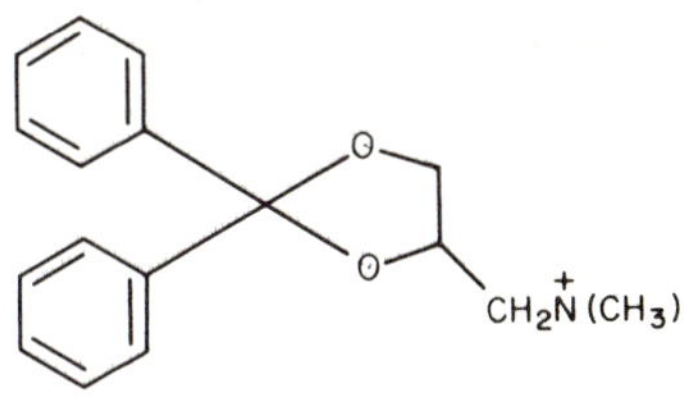

42

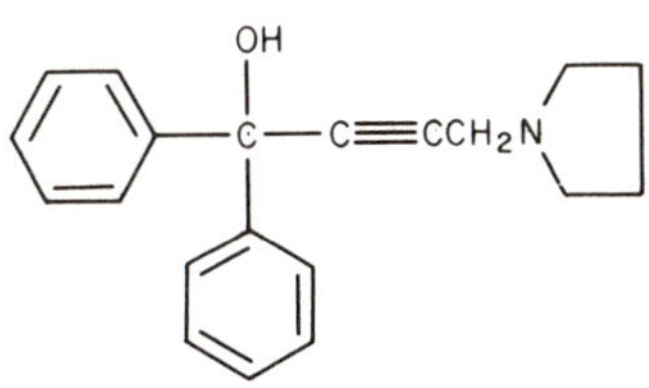

43

This relationship between the structures of agonists and antagonists, taken together with the fact that a parallel shift in the dose-response curve of acetylcholine on the isolated guinea-pig ileum occurs in the presence of an antiacetylcholine drug, suggests strongly that such drugs are acting as competitive antagonists. This implies at least one common point of interaction at the receptor for agonists and antagonists. The most obvious point for such a common interaction would seem to be site 3 on the proposed receptor outlined above, where the quaternary group of muscarinic agents and the amine group of antiacetylcholine drugs would be assumed to interact.

The role of the positively charged cationic group in the interaction of drugs with the muscarinic receptor was discussed by Burgen (1965). He concluded that interaction of the cationic end of acetylcholine, and presumably other agonists, with the appropriate receptor site played a dominant role in the production of agonist activity. With antagonists, on the other hand, the large substituents at the acyl end of the molecule were involved in the most important drug-receptor interactions and this resulted in a less close contact between the cationic end of the molecule and the anionic group of the receptor. This idea certainly goes some way toward explaining the difference between agonists and antagonists.

A somewhat different view had been put earlier by Barlow, Scott and Stephenson (1963). They tested parallel series of compounds related to acetylcholine for muscarinic agonist or antagonist activity. One series, mainly agonists, had chains of five carbon and/or oxygen atoms attached to an onium group. The other series, of antagonists, had the terminal methyl group replaced by Ph_2CH or $Ph_2C(OH)$. Methyl was replaced by ethyl in the onium group and the effects on affinity were found to be similar in the different series of antagonists. It was suggested that similar effects on affinity would be achieved by similar structural changes in corresponding series of agonists. The general conclusion, then, differed from that of Burgen in that it implied that the affinity of the cationic end of the molecule was the same in agonists and antagonists and that the effects on affinity of altering the nature of the onium group were not very dependent on the nature of the rest of the molecule.

In a later paper Barlow and Stephenson and co-workers (Abramson, *et al.*, 1969) studied various series of analogues of acetylcholine. Measurements were made of the affinity of the compounds for the muscarinic receptor and in these series it was found that changes in affinity as the onium group was modified were not entirely independent of changes in the rest of the molecule. This seems to be at variance with the previous postulate of Barlow, Scott and Stephenson (1963) but it is very difficult to make any clearcut interpretation of the results. As Abramson *et al.* pointed out, the introduction of an additional group into the molecule does not only affect the binding of the drug to the receptor, it can also effect a realignment of the rest of the molecule which may cause the fit to be less good and thus tend to reduce binding. Other factors may also be involved and it is very difficult to assess their overall effect on affinity, especially when they do not all affect it in the same direction. Abramson *et al.*, concluded that their results did not support Burgen's view that the separation between the cationic head of the drug and the anionic group in the receptor was greater in compounds with high affinity, that is, in potent antagonists.

In view of this controversy it seems that serious consideration should be given to another possibility, that muscarinic agonist and antagonist drugs do not interact with the same receptor. Such a suggestion was made by Goldstein, Aronow and Kalman (1968) who stated that an apparent competitive antagonism could result if the receptors were quite distinct but that the presence of an antagonist near to the acetylcholine receptor could modify the latter in such a way that the affinity of an agonist for it would be reduced. Some qualified support for this kind of view has been given by Brimblecombe, Green and Inch (1970) who have made a comparison of stereochemical requirements of muscarinic and antiacetylcholine drugs. This was based to a large degree on a study of structure-activity relationships in antiacetylcholine dioxolans by Brimblecombe and Inch (1970). All these compounds were based on the 1,3-dioxolan structure (see formula 29) and they were tested for antiacetylcholine activity *in vivo* by measuring their potency in producing mydriasis and in antagonising oxotremorine effects in mice. Their affinity constants (K_B) were measured *in vitro* by their ability to antagonise acetylcholine-induced contractions of the isolated guinea-pig ileum.

In an initial series of experiments pairs of *cis* and *trans*-2-substituted-4-dimethylamino-methyl-1,3-dioxolan hydrochlorides and methiodides in the racemic form and of previously established geometry were studied. The hydrochlorides were weak antiacetylcholine agents; the results obtained with the methiodides are given in Table 2.4.

TABLE 2.4

Pharmacological results for 1,3-Dioxolans of general structure

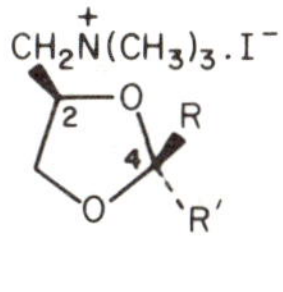

(From Brimblecombe and Inch, 1970)

R	R′	log K (Affinity constant guinea pig ileum)	ED50 μmole/kg with 95% limits for antagonism of oxo-tremorine-induced salivation in mice	Mydriatic effect in mice. Potency relative to atro-pine with 95% limits
CH_2Ph	Ph	8.05	13.8(7.5–42.3)	0.21(0.17–0.27)
Ph	CH_2Ph	8.35	6.1(1.0–12.0)	0.62(0.47–0.86)
Ph	Ph	8.02	15.5(6.9–28.2)	0.50(0.39–0.64)
C_6H_{11}	Ph	8.41	15.0(8.6–27.1)	0.61(0.44–0.94)
Ph	C_6H_{11}	8.34	6.2(4.3–9.0)	0.98(0.79–1.28)
CH_3	Ph	5.30	> 100	< 0.01
CH_3	CH_2Ph	4	> 100	< 0.01
CH_2Ph	CH_3	5.11	> 100	< 0.01
H	Ph	4.52	> 100	< 0.01
Ph	H	4.31	> 100	< 0.01

As a result *cis*- and *trans*-2-cyclohexyl-4-dimethylamino-methyl-2-phenyl-1,3-dioxolan methiodides were prepared in optically pure forms. The results for these optically pure isomers and the racemates are given in Table 2.5.

TABLE 2.5

Pharmacological results for racemates and optical isomers of cis *and* trans-*2-cyclohexyl-4-dimethylaminomethyl-2-phenyl-1,3-dioxolan methiodides*

(From Brimblecombe and Inch 1970)

Configuration	log K	ED50 for antagonism of oxotremorine-induced salivation	Mydriatic effect Potency relative to atropine
trans-racemate	8.41	15.0(8.6–27.1)	0.61(0.44–0.94)
D-*trans* (2S,4S)	8.56	4.2(2.4–7.3)	1.12(0.85–1.64)
L-*trans* (2R,4R)	8.24	20.2(15.3–26.7)	0.29(0.22–0.38)
cis-racemate	8.34	6.2(4.3–9.0)	0.98(0.79–1.28)
D-*cis* (2R,4S)	8.34	16.6(11.5–23.9)	0.74(0.56–1.07)
L-*cis* (2S,4R)	8.73	2.9(2.1–4.1)	2.29(1.79–3.02)

It is clear from the results in Table 2.4 that compounds of this type having two bulky substituents at C-2 have antiacetylcholine potencies not very different from that of atropine. Compounds with only one bulky substituent on this carbon atom are, however, only weak antiacetylcholine agents.

The first point to emerge from an examination of the results in Table 2.5 is that the assay procedures used seem to give results which are internally consistent in that the values for the racemates lie between the values for the corresponding optical isomers. Secondly, it is apparent that differences in activity between the compounds are not very great but in descending order of potency they can be arranged thus:

$$\text{L-}cis > \text{D-}trans > \text{D-}cis > \text{L-}trans$$

With this information, and with knowledge of the steric factors which can contribute to antiacetylcholine activity, it is possible to decide which factors play the major role. The three main factors are the absolute configuration at C-2, the absolute configuration at C-4 and the geometrical relationship between the substituents at C-2 and at C-4.

The L-*cis* and the D-*trans* compounds differ in configuration at C-4 and have a different geometrical relationship between C-2 and C-4 so it seems apparent that it is the absolute configuration at the acetal carbon which contributes most to antiacetylcholine activity. This is in contrast to the situation in the corresponding cholinergic dioxolans where it is the configuration at C-4 and not at the acetal carbon which affects activity (Belleau and Lavoie, 1968).

This difference between the stereochemical requirements for agonist and antagonist activity in corresponding compounds led Brimblecombe, Green and Inch (1970) to consider other analogous situations. Ellenbroek, Nivard, Van Rossum and Ariens (1965) studied esters of 2-cyclohexyl-2-hydroxy-2-phenylacetic acid which are structurally related to acetylcholine. They found that in antiacetylcholine drugs derived from this acid the potency was critically dependent on the configuration of the benzylic carbon atom with the R enantiomer being more active than the S enantiomer by two orders of magnitude (see formula 44).

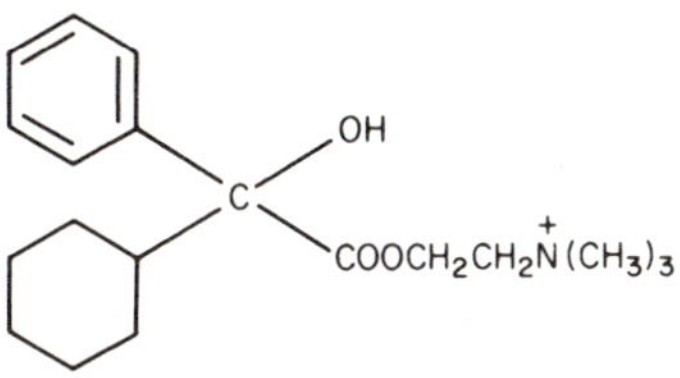

44

$CH_3COOCHR'CHR\overset{+}{N}(CH_3)_3$

45

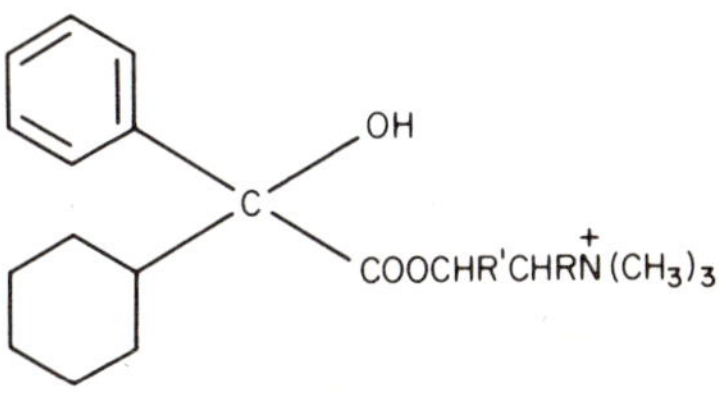

46

It is possible to compare the activities of agonist esters of acetic acid and antagonist esters of R(−)2-cyclohexyl-2-hydroxy-2-phenylacetic acid thus (Brimblecombe, Green and Inch, 1970):

(1) A reduction in agonist activity occurs when any of the N-methyl substituents in compound 45 are replaced by other alkyl groups. In compound 46 on the other hand, the nature of the *N* substituents can be varied considerably without appreciably affecting antiacetylcholine potency. Only quaternary compounds are potent agonists whereas for antagonist activity the nitrogen can be tertiary or quaternary (Abood, 1968).

(2) Substitution of an α-methyl group in compound 45 (R = CH_3) to give α-methylcholine produces a considerable reduction in muscarinic

potency. In this compound potency is dependent on the absolute configuration of the methyl-substituted carbon with the R enantiomer being eight times more active than the S enantiomer (Beckett, Harper and Clitherow, 1963). Substitution with an α-methyl group in compound 46 enhances, rather than reduces, antiacetylcholine potency and the configuration of the methyl-substituted carbon atom does not influence this potency.

(3) The S enantiomer of β-methylcholine (see formula 45, R′ = CH_3) is much more active than the R enantiomer (Beckett, Harper and Clitherow, 1963). Methyl substitution of the β-carbon of compound 46 reduces activity and gives a compound in which the absolute configuration of the methyl-substituted carbon atom has little influence on potency.

(4) A considerable reduction in muscarinic potency results from replacement of the alcoholic oxygen in acetylcholine with sulphur. Similar replacement in antiacetylcholine compounds has little effect on potency (Barlow, 1964).

These examples taken with the results from the dioxolans given above serve to show that the stereochemical requirements for high potency differ widely between agonists and the corresponding antagonists. One possible explanation of this would seem to be that the two types of drugs do not, as has been generally assumed, interact with the same receptor. It is very difficult, however to envisage any satisfactory biological reason for the presence of possible "antagonist receptors" and clearly this explanation should not be fully accepted, or completely rejected, without a good deal more experimental work.

As well as differences between structure-activity relationships in corresponding series of agonists and antagonists such differences also exist between different series of antagonists, some of them fairly closely related structurally. For example, Bebbington and Brimblecombe (1969) compared the antiacetylcholine activities of corresponding series of acetylenic carbinols and acetylenic glycollates. The results are given in Table 2.6.

A general point to be made concerning these results is that the compounds are somewhat less active than the corresponding non-acetylenic glycollates examined by Abood (1968). This may well be related to the rigidity imposed on the molecule by the acetylenic bond presumably resulting in a less good fit to the receptor but this does not really provide evidence for or against the concept of there being separate receptors for agonists and antagonists. The same applies to the differences which are apparent between the series of carbinols and glycollates.

It is apparent that a good deal remains to be discussed concerning the

TABLE 2.6

A comparison of the antiacetylcholine activities of acetylenic carbinols and acetylenic glycollates

From Bebbington and Brimblecombe (1969)

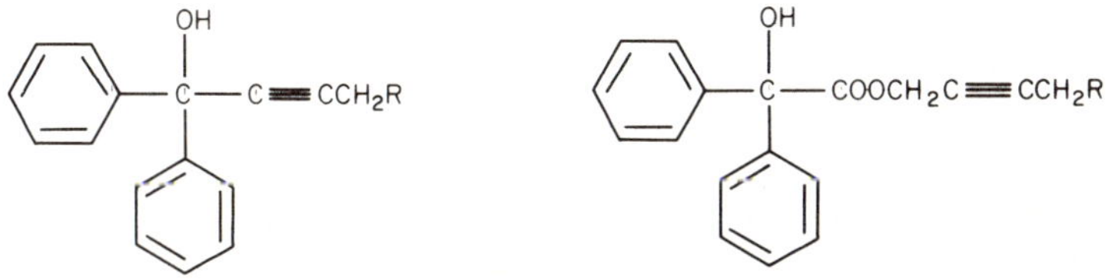

R	pA2	ED50 (μmole/kg) for antagonism of oxotremorine induced salivation in mice	Mydriatic potency relative to atropine in mice	pA2	ED50 (μmole/kg) for antagonism of oxotremorine induced salivation in mice	Mydriatic potency relative to atropine in mice
$N(CH_3)_2$	7.8	22.1	0.02	7.95	5.7	0.34
N	8.1	10.2	0.11	7.45	29.9	0.05
N	8.1	8.8	0.60	7.10	100	0.02
N	8.1	100	0.02	7.00	100	0.01

interactions of antiacetylcholine drugs with their receptor. For the remainder of this monograph, in particular in the Chapter dealing with central activities of such drugs, the assumption will be made that they interact with the same receptor as agonists and act as classical competitive antagonists. That this assumption may not be correct has been indicated above but the methods of analysis of drug effects in the central nervous system are not nearly precise enough to enable such fine distinctions to be made.

References

Abood, L. G. (1968). The psychotomimetic glycolate esters. *Drugs affecting the Central Nervous System.* (ed. A. Burger) Marcel Dekker, New York, pp. 127–167

Abramson, F. B., Barlow, R. B., Mustafa, M. G. and Stephenson, R. P.

(1969). Relationships between chemical structure and affinity for acetylcholine receptors. *Br. J. Pharmac.*, **37**, 207–233

Ariens, E. J. (1966). Receptor theory and structure-action relationships. *Adv. Drug. Res.*, **3**, 235–285

Barlow, R. B. (1964). *Introduction to Chemical Pharmacology*, Methuen & Co, Ltd., London

Barlow, R. B., Scott, K. A. and Stephenson, R. P. (1963). An attempt to study the effects of chemical structure on the affinity and efficacy of compounds related to acetylcholine. *Br. J. Pharmac. Chemother.*, **21**, 509–522

Bebbington, A. and Brimblecombe, R. W. (1965). Muscarinic receptors in the peripheral and central nervous system. *Adv. Drug. Res.*, **2**, 143–172

Bebbington, A. and Brimblecombe, R. W. (1969). Actions of some toxic substances (psychotomimetics) on the central nervous system. *Br. Med. Bull.*, **25**, 293–298

Bebbington, A., Brimblecombe, R. W. and Shakeshaft, D. (1966). The central and peripheral activity of acetylenic amines related to oxotremorine. *Br. J. Pharmac. Chemother.*, **26**, 56–57

Beckett, A. H., Harper, N. J., Clitherow, J. W. and Lesser, E. (1961). Muscarinic receptors. *Nature (Lond.)*, **189**, 671–673

Beckett, A. H., Harper, N. J. and Clitherow, J. W. (1963). The importance of stereoisomerism in muscarinic activity. *J. Pharm. Pharmac.*, **15**, 362–371

Belleau, B. and Lavoie, J. L. (1968). A biological basis of ligand-induced activation of excitable membranes and associated enzymes. A thermodynamic study using acetylcholinesterase as a model receptor. *Can. J. Biochem.*, **46**, 1397–1409

Belleau, B. and Puranen, J. (1963). Stereochemistry of the interaction of enantiomeric 1,3-dioxolane analogs of muscarone with cholinergic receptors. *J. Med. Chem.*, **6**, 325–328

Brimblecombe, R. W. and Inch, T. D. (1970). Structure-activity relations for anticholinergic dioxolans. *J. Pharm. Pharmac.*, **22**, 881–888

Brimblecombe, R. W., Green, D. M. and Inch, T. D. (1970). A comparison of the stereochemical requirements of cholinergic and anticholinergic drugs. *J. Pharm. Pharmac.*, **22**, 951–953

Brown, B. B. and Werner, H. W. (1949). The parasympathetic depressant and antihistaminic action of substituted 1,3-dioxolanes. *J. Pharm. exp. Ther.*, **97**, 157–170

Burgen, A. S. V. (1965). The role of ionic interaction at the muscarinic receptor. *Br. J. Pharmac. Chemother.*, **25**, 4–17

Canepa, F. G. (1965). Relations between acetylcholine, choline and betaine chains in aqueous solution. *Nature, Lond.*, **207**, 1152–1154

Cho, A. K., Haslett, W. L. and Jenden, D. J. (1962). The peripheral actions of oxotremorine, a metabolite of tremorine. *J. Pharmac. exp. Ther.*, **138**, 249–257

Dahlbom, R., Karlen, B., Ramsby, S., Kraft, I. and Mollberg, R. (1964). Acetylene compounds with potential pharmacological value. IV. 1,1-di-substituted-4-dialkylamino-2-butyn-1-ols. *Acta. pharm. suecica.*, **1**, 237–246

Ellenbroek, B. W. J., Nivard, R. J. F., Van Rossum, J. M. and Ariens, E. J. (1965). Absolute configuration and parasympathetic action, pharmacodynamics of enantiomorphic and diastereoisomeric esters of methylcholine. *J. Pharm. Pharmac.*, **17**, 393–404

Fodor, G. and Csepreghy, G. (1961). The stereochemistry of the tropane alkaloids. Pt XIV. The absolute configuration of (–)–tropic acid, hyoscyamine and hyoscine. *J. Chem. Soc.*, 3222–3227

Fourneau, E., Bovet, D., Bovet, F. and Montezin, G. (1944). Activité acetylcholinique particulièrement intense dans une nouvelle série d'aminoacetals de polyols. *Bull. Soc. Chem. Biol.*, **26**, 516–527

Gaddum, J. H. (1957). Drug antagonism. *Pharmac. Rev.*, **9**, 211–218

Goldstein, A., Aronow, L. and Kalman, S. (1968). *Principles of Drug Action.* Harper & Row, New York

Gyermek, L. and Nador, K. (1957). The pharmacology of tropane compounds in relation to their steric structure. *J. Pharm. Pharmac.*, **9**, 209–229

Gyermek, L. and Unna, K. R. (1958). Relations of structures of synthetic muscarines and muscarones to their pharmacological action. *Proc. Soc. exp. Biol. N.Y.*, **98**, 882–885

Hardeggar, E. and Lohse, F. (1957). Muscarine. VII. Synthesis and absolute configuration of muscarine. *Helv. Chem. Acta.*, **40**, 2383–2389

Ing, H. R. (1949). The structure-action relationships of the choline group. *Science*, **109**, 264–266

Ing, H. R. (1955). *The Alkaloids*, Vol. 5. (ed. Manske & Holmers) Academic Press, New York, p 243

Ing, H. R., Kordik, P. and Tudor-Williams, D. P. H. (1952). Studies on the structure-action relationships of the choline group. *Br. J. Pharmac. Chemother.*, **7**, 103–116

Jacob, J., Marszak, I., Bardisa, L., Marszak-Fleury, A. and Epsztein, R. (1952). Relation of structure to activity in C_2H_2, C_2H_4 and saturated derivatives of choline and acetylcholine. *Arch. int. Pharmacodyn.*, **91**, 103–110

Kier, L. B. (1967). Molecular orbital calculation of preferred conformations of acetylcholine, muscarine and muscarone. *Mol. Pharmac.*, **3**, 487–494

Kier, L. B. (1970a). Receptor mapping using molecular orbital theory. *Fundamental Concepts in Drug-Receptor Interactions*. (eds. J. F. Danielli, J. F. Moran and D. J. Triggle) Academic Press, New York, pp 15–45

Kier, L. B. (1970b). Molecular orbital conformation of oxotremorine and a comparison with the muscarinic pattern. *J. Pharmaceut. Sci.*, **59**, 112–114

Koshland, E. E. (1959). Mechanisms of transfer enzymes. *The Enzymes*, Vol. I. (eds. P. D. Boyer, H. Lardy and K. Myrback) Academic Press, New York, pp 305–346

Kuznetzov, S. G. and Golikov, S. N. (1962). *Synthetic Atropine-like Substances*. State Publicity House of Medical Literature, Leningrad. English trans: U.S. Dept. Comm. J.P.R.S. 19757

Martin-Smith, M., Smail, G. A. and Stenlake, J. B. (1967). The possible role of conformational isomerism in the biological actions of acetylcholine. *J. Pharm. Pharmac.*, **19**, 561–589

Schild, H. O. (1947). pA, a new scale for the measurement of drug antagonism. *Br. J. Pharmac. Chemother.*, **2**, 189–206

Stehle, R. L., Melville, K. F. and Oldham, F. K. (1936). Choline as a factor in the elaboration of adrenaline. *J. Pharmac. exp. Ther.*, **56**, 473–481

Stephenson, R. P. (1956). A modification of receptor theory. *Br. J. Pharmac. Chemother.*, **11**, 379–393

Van Rossum, J. M. and Ariens, E. J. (1959). Pharmacodynamics of parasympathetic drugs. *Arch. int. Pharmacodyn.*, **118**, 418–446

Waser, P. G. (1961). Chemistry and pharmacology of muscarine, muscarone and related compounds. *Pharmac. Rev.*, **13**, 465–515

3 Drug Actions at Peripheral Nicotinic Sites

Dale noted in 1914 that the actions of acetylcholine at the neuromuscular junction and at the autonomic ganglia were mimicked closely by nicotine. Hence these sites have become known as "nicotinic" and drugs acting at them as nicotinic agonists or antagonists. Medicinally, ganglion blocking agents are used in the treatment of hypertension, tending to reduce the vasoconstriction brought about by tonic sympathetic impulses. Such drugs have also found some use in the treatment of gastric and duodenal ulcers when they reduce gastric and duodenal mobility and diminish gastric acid secretion. The main difficulty with the clinical use of ganglion blockers lies in their lack of specificity. For antihypertensive therapy blockade of sympathetic ganglia is required, whereas in the management of gastric and duodenal ulcers a parasympathetic ganglion blockade is necessary. Unfortunately drugs have not been found with specific actions on either type of ganglion.

In vitro methods for the study of ganglion-stimulants and ganglion-blockers are not entirely satisfactory for a number of reasons. The ganglia normally used are parasympathetic and lie in close juxtaposition to the tissues which are being innervated, the result being that effects on the ganglia are not easily distinguished from those on the tissue itself. The isolated guinea-pig ileum has been used commonly; ganglion stimulants produce a contraction of the tissue which is blocked by ganglion blockers. Alternatively, the method of Trendelenberg has been employed in which the peristaltic reflex of guinea-pig ileum can be recorded. This method is, however, difficult to quantitate for use in accurate assays. A preparation which shows some promise is the isolated hypogastric nerve-vas deferens of the guinea-pig first described by Hucovics in 1961. A later study of the preparation by Bentley and Sabine (1963) revealed that the contractions of the guinea-pig first described by Hucović in 1961. A later study of the blocked by ganglion-blocking agents but that these drugs were without effect on the contractions elicited by transmural stimulation although these contractions were blocked by sympathetic blockers. This, then, constitutes a preparation in which ganglion blocking activity can be studied specifically and it has recently been used successfully by Cooper, Green,

Rickard and Thompson (1971) to assay the activity of some diastereoisomeric dimethylaminobornyl acetates and their methiodides.

The most commonly employed *in vivo* method for studying drug actions on ganglia involves the use of the cat superior cervical ganglion *in situ*. Contraction of the nictitating membrane is normally used as the response to ganglion stimulation although post-ganglionic action potentials have sometimes been recorded. Ideally the drugs should be applied as near to the ganglion as possible by close intra-arterial injection into the external carotid or lingual artery. The hypertensive response in anaesthetised, atropinised cats or rats has also been used but the mechanism of this response is complicated and is not necessarily mediated solely through sympathetic ganglion stimulation.

Neuromuscular blocking agents are used in surgery in conjunction with anaesthetics to obtain muscle relaxation. Similarly, in psychiatry these agents are used to prevent excess motor activity and consequent danger of limb fractures or dislocations during electroconvulsive therapy for depression.

In vitro preparations for studying the actions of drugs at the neuromuscular junction include isolated leech, frog or avian muscles all of which respond to nicotinic agonists by a slow contracture. Various isolated nerve-muscle preparations are used to study neuromuscular blocking agents. These include the rat phrenic nerve-diaphragm preparation and the chick biventer-cervicis. Muscle contractions can be evoked by direct or indirect (that is, via the nerve) stimulation so that the site of action of a drug can be defined as being at the neuromuscular junction or on the contractile mechanism of the muscle itself. *In vivo* nerve-muscle preparations are also commonly used with the tibialis, soleus or flexor hallucis longus muscles of anaesthetised cats or rabbits being stimulated via the appropriate motor nerve. A qualitative whole animal test using chickens was described by Buttle and Zaimis (1949) to distinguish between non-depolorising or curare-like neuromuscular blocking agents and depolorising blockers. The former drugs, when injected into a wing vein, produce a plastic paralysis, with consequent head drop while the latter cause a spastic paralysis with head retraction.

No attempt will be made in this monograph to review the whole field of drug action at autonomic ganglia and the neuromuscular junction; for this purpose reference should be made to Barlow (1964), Volle and Koelle (1970), Koelle (1970) and to various specialist review articles many of which are referred to in the references just listed.

Here three topics, of particular interest to the author, are considered. They are:

(1) A study of the action of some ganglion stimulants in an attempt

to ascertain whether their actions result from direct interaction with post-synaptic receptors or from a pre-synaptic action resulting in release of acetylcholine.

(2) An account of the actions of some antimuscarinic drugs at the neuromuscular junction, the traditional site of antinicotinic drugs. This is of particular relevance to the discussion, in the next Chapter, of the role of antimuscarinic drugs in the treatment of poisoning with anticholinesterase agents.

(3) A discussion of whether the differences which clearly exist between drug actions at the autonomic ganglia and at the neuromuscular junction can be accounted for simply by differences in receptor distribution or whether fundamental qualitative differences exist between the receptors at these two cholinoceptive sites. This is of particular relevance to the discussion later in this monograph of the possible existence of central nicotinic receptors. If nicotinic receptors are indeed of two types—the "ganglion type" and the "neuromuscular junction type" then it will be necessary, assuming that such receptors are present in the central nervous system, to ascertain which type of peripheral receptor they most resemble.

3.1 Actions of some ganglion-stimulant drugs

A historical account of drug-actions on autonomic ganglia is given by Volle (1966) and in the same monograph it is made quite clear that the original concept of all ganglionic transmission being mediated through acetylcholine receptors of the nicotinic type is no longer tenable. Volle and Koelle (1970) pointed out that this form of transmission should still be regarded as the primary pathway but that there are at least two other secondary pathways for the transmission of excitatory and inhibitory impulses. Unlike the primary pathway these are not blocked by hexamethonium. In sympathetic ganglia it is possible to show that acetylcholine elicits a triphasic postsynaptic response with an initial excitatory postsynaptic potential (EPSP) followed in turn by an inhibitory postsynaptic potential (IPSP) and a late EPSP. The initial EPSP is blocked by hexamethonium and so is assumed to initiate the primary transmission process. The late EPSP is blocked by atropine but is unaffected or enhanced by hexamethonium; it may be involved in an alternative pathway if the primary route of transmission is blocked. The IPSP is depressed by atropine and unaffected by hexamethonium but is prevented by α-adrenoceptor blocking agents suggesting the involvement of adrenaline, probably as an inhibitory transmitter, in sympathetic ganglia.

It was demonstrated by Koelle and Koelle in 1959 that essentially all the functional acetylcholinesterase in autonomic ganglia of the cat is presynaptic. The physiological function of this pre-synaptic enzyme was subsequently investigated by Volle and Koelle (1961) who concluded that its major role was to protect the pre-synaptic nerve terminals from endogenously liberated acetylcholine. These pre-synaptic terminals were considered to be the primary site of action of carbachol (see formula 24) which caused release of acetylcholine which in turn interacted with, and stimulated, post-synaptic receptors. The evidence for this suggestion was based on the finding that the mean threshold dose of carbachol for ganglion stimulation was twenty-six times greater in the chronically denervated ganglion than in the normal ganglion. This finding has subsequently been disputed by other workers. In particular, Brimblecombe and Sutton (1968) investigated the ganglion-stimulating effects of some amino-acid esters, the detailed pharmacology of which has been described by Barrass, Brimblecombe, Parkes and Rich (1968). These drugs, (see general formula 47), were predominantly nicotinic but also possessed some muscarinic activity and it was because of this dual activity that their actions on the superior cervical ganglion of the cat were studied.

$$R_3\overset{+}{N}(CH_2)_nCO_2R'.X^-$$

47

The esters, when administered by close intra-arterial injection, produced effects (contraction of the nictitating membrane and asynchronous discharges in post-ganglionic nerve fibres) which were blocked by hexamethonium and reduced by atropine. These effects were identical to those produced by carbachol and were unaffected by pretreatment with the anticholinesterase drug dyflos. The reduction in sensitivity of the chronically denervated ganglion, reported by Volle and Koelle (1961) was not found. On the contrary, the denervated ganglion was, if anything, slightly more sensitive to the amino-acid esters and to carbachol. Similarly, Brown (1968) reported a series of experiments in which denervation did not reduce the sensitivity of the cat superior cervical ganglion to a series of drugs, including carbachol. Later, Brown, Halliwell, Jones and Quilliam (1970) made a direct test of the hypothesis that acetylcholine itself can act on pre-synaptic terminals to liberate further acetylcholine which in turn stimulates post-synaptic receptors. These workers labelled ganglionic acetylcholine with tritium and then observed the effect on release of 3H-acetylcholine of injection of unlabelled acetylcholine. In fact, neither acetylcholine nor other ganglion-stimulants including carbachol, tetramethylammonium and nicotine had any effect on acetylcholine release.

These findings then, seem to raise serious doubts as to the presence of presynaptic ganglionic cholinoceptive sites. Another relevant finding (Volle and Koelle, 1961; Brimblecombe and Sutton, 1968) is that the response to carbachol, unlike that to acetylcholine, is not potentiated by an anticholinesterase drug as might be expected if release of acetylcholine is involved. Volle and Koelle (1961) offer the explanation that, because all the acetylcholinesterase is confined to the pre-synaptic terminals, it might be retained in this area and so be effective in limiting the pre-synaptic, but not the post-synaptic actions of acetylcholine.

The evidence is clearly conflicting but enough doubt seems to have been created to necessitate a careful re-appraisal of the possibility of the existence of pre-synaptic cholinoceptive sites and indeed of the general role of the pre-synaptic acetylcholinesterase.

3.2 Neuromuscular effects of some antimuscarinic drugs

While the main site of action of antimuscarinic or antiacetylcholine drugs is at the postganglionic parasympathetic endings it has been known for some time that many of these drugs are also capable of exerting actions at other sites including the neuromuscular junction and the skeletal muscle fibre. The earliest relevant observation appears to have been that of Botkin (1862) who noted that large doses of atropine sulphate caused paralysis in experimental animals and suggested that this was due to a peripheral action of the drug. Subsequently, a number of workers confirmed that atropine produced a neuromuscular blockade (Gushny, 1903; Haffner, 1914; Dale and Gaddum, 1930; Abdon, 1945). This became known as the "curare-like" action of atropine but in a study on the isolated rat phrenic nerve-diaphragm preparation Bulbring (1946) found that atropine differed from curarine. Like the latter it abolished the depressant action of rapid motor nerve stimulation but, on the other hand, it did not abolish the depression caused by the addition of acetylcholine to the tissue bath. Bulbring found that the action of atropine was similar to that of procaine and it seemed to have a significant pre-synaptic action as well as a curare-like action. Procaine appears to reduce the output of acetylcholine from motor nerve endings (Harvey, 1939; Jaco and Wood, 1944). By analogy atropine might be assumed to have the same action. A somewhat different conclusion to that of Bulbring was reached by Beranek and Vyskocil (1967). They compared the actions of tubocurarine and atropine on the rat diaphragm and showed that both drugs reduced the magnitude of the end-plate potentials. The concentrations of atropine required for this action was two thousand times greater than that of tubocurarine but the authors concluded that neither drug had any significant pre-synaptic action in concentrations which markedly reduced the amplitude of the end-plate potentials. Later Beranek and Vyskocil (1968) studied the actions of the same two drugs on the frog sartorius neuromuscular junction and found that they both reduced the amplitude of intracellularly recorded end-plate potentials and miniature end-plate potentials with the concentration of atropine required for this effect being one hundred times greater than that of tubocurarine. The potency of atropine was approximately equal on the rat diaphragm preparation and the frog sartorius preparation but the potency of tubocurarine was much lower on the frog muscle preparation.

As well as this neuromuscular blocking action atropine, in lower concentrations, has a potentiating effect on the twitches produced by indirect stimulation of the isolated rat phrenic nerve-diaphragm preparation (Bulbring, 1946; Dutta, 1949; Segawa, Kojima and Takagi, 1967). A similar dose-dependent biphasic action of atropine *in vivo* was demonstrated by Brimblecombe and Everett (1969, 1970). Using fast-twitch (flexor hallucis longus, FHL) and slow-twitch (soleus) muscles of the anaesthetized cat they were able to show that doses of less than 8 mg given by close intra-arterial injection into a small branch of the femoral artery potentiated twitches of both muscles. Higher doses depressed both muscles. Very similar effects were observed with 2-diethylaminoethyl-phenylcyclopentane-1-carboxylate (Caramiphen, see formula 48) but the effects of hyoscine were variable in that while it produced only depression of soleus it sometimes produced potentiation of FHL followed by depression; in other experiments only depression was seen. Attention in this series of experiments was focused on another antiacetylcholine drug, N-ethyl-2-pyrrolidylmethylcyclopentylphenyl glycollate (PMCG, see formula 49).

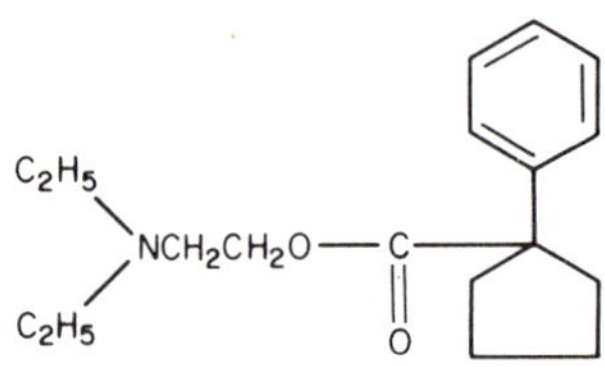

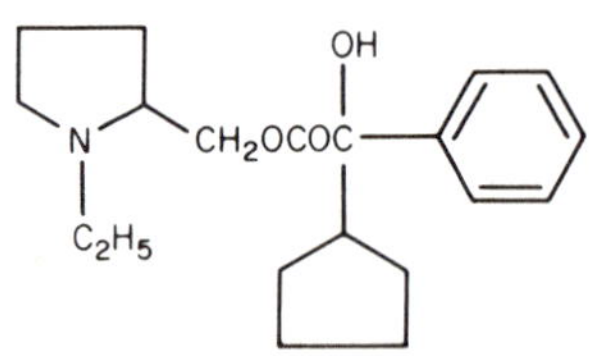

49

This drug, which is the major component (70%) of Ditran, has both peripheral and central atropine-like properties (Abood and Biel, 1962). Brimblecombe and Green (1968) found it to be about four times less active than atropine peripherally and five times more active centrally. Abood and Biel (1962) reported that, at a concentration of 10^{-6}M, the drug increased by up to 50%, isometric twitches of the isolated, indirectly-stimulated, frog sartorius muscle in Ringer's solution but inhibited spontaneous twitches of the muscle in a calcium-free solution. These authors suggested that PMCG affected muscle depolarization by interfering with the movement of ions, particularly Mg^{++} and Ca^{++}.

In experiments using the rat phrenic nerve-diaphragm preparation it has been found (Brimblecombe and Everett, 1969; Brimblecombe, Green, Stratton and Thompson, 1970) that PMCG increases muscle twitches by up to 47 % at a bath concentration of 10 μg/ml. In *in vivo* experiments in cats (Brimblecombe and Everett, 1969, 1970) twitches of both FHL and soleus were markedly potentiated by intra-arterial doses of PMCG of 0.25 to 10 mg. Maximum potentiation in FHL was 200 % and in soleus 70 %. The increases in twitch tension were characterised by increase in the maximum rate of rise of tension and in the time to reach the peak tension. At doses above 10 mg there was usually an abrupt change to pronounced depression of the twitch. With soleus this depression was more marked and more persistent than with FHL. During potentiation no repetitive firing of the muscle fibres could be recorded from surface recording electrodes and, in any case, PMCG has only very weak anti-cholinesterase activity (pI50 for acetylcholinesterase = 1.5, and for cholinesterase = 4.0) so this can be ruled out as being of any significance in the potentiating action. Additionally, it was found that effects of PMCG on fully curarized, directly stimulated, muscles were essentially the same as those on indirectly stimulated muscles although depression was only seen after high doses (about 50 mg).

Tetanic responses of both FHL and soleus were also modified by PMCG. The nature of the modification varied with dose and with frequency of stimulation but essentially the findings were that doses of PMCG which potentiated twitch also potentiated low frequency tetani (5–30 Hz for soleus, 10–50 Hz for FHL). There was both augmentation of the units of contraction and an increase in the degree of fusion at a given frequency. These doses depressed and caused a non-maintenance of higher frequency tetani (80 Hz and above for soleus, 150 Hz and above for FHL). Tetani of intermediate frequency (40–60 Hz for soleus, 80–100 Hz for FHL) were little affected. The other drugs referred to above, atropine, caramiphen and hyoscine, affected tetanic responses in essentially the same manner and on the rat phrenic nerve-diaphragm preparation a similar picture was seen with potentiation of responses to 5 and 30 Hz stimulation but a partial block of the Wedensky type with a frequency of stimulation of 50 Hz.

The fact that, in the cat experiments, PMCG and other drugs caused potentiation of twitches in the directly, as well as the indirectly, stimulated muscle makes it certain that their main site of action is beyond the neuromuscular junction, that is, on the muscle fibres themselves. This conclusion would be compatible with the suggestions of Abood

(1968) that various glycollic acid esters, all of which he unfortunately terms PMCG, interact with ATP, Ca^{++} and lipophilic components in the muscle membrane to interfere with the calcium-dependent excitation-contraction coupling of the muscle. The drugs were also considered to be capable of substituting for Ca^{++} in restoring the resting potential.

The findings of Brimblecombe and Everett (1969, 1970) also seem to be in general agreement with those of Beranek and Vyskocil (1967) in that the latter authors concluded that the main site of action of atropine was post-synaptic. Nevertheless, the action of PMCG, and probably of atropine also, is clearly a complex one. It seems to have a rather more typical curare-like action in addition to the action described above. Evidence for this lies in the fact that in the cats, depression of indirectly stimulated muscles occurred at doses of PMCG lower than those causing depression of directly stimulated muscles. There are, however, differences between PMCG and curare. Typically, after curare repetitive stimulation produced a rapid decline in tetanic tension which then levelled off. After PMCG, on the other hand, there was a gradual, progressive fall in tetanic tension.

There may be other factors contributing to the neuromuscular actions of these drugs. Local anaesthetics such as procaine cause increase in twitch tension and prolongation of the negative after-potential (Lammers and Ritchie, 1955; Falk, 1961) and recently it has been shown (Brimblecombe, unpublished results) that PMCG has appreciable local anaesthetic activity. Caffeine also produces twitch augmentation (Brimblecombe and Webb, 1973). Clearly more studies are required to elucidate the precise mode of action of these drugs. In the meantime two points are worth noting. Firstly, it should always be borne in mind that drugs are rarely completely specific in their actions. In this particular instance while PMCG and atropine are predominantly antimuscarinic, under some circumstances their actions at the neuromuscular junction may become of significance. This is relevant to the second point which is concerned with the role of these drugs in the treatment of poisoning with anticholinesterase agents. Their value lies mainly in their effects in antagonising the actions of accumulated acetylcholine at muscarinic sites but their curare-like actions might also contribute by antagonising the actions of acetylcholine at the neuromuscular junction. This matter is discussed more fully in the next Chapter.

3.3 Ganglionic and neuromuscular junction receptors

There are many examples of drugs having different levels of activity at the neuromuscular junction and at autonomic ganglia. No attempt will be made here to review the whole field; reference should be made to a standard text, for example Barlow (1964). One of the most striking examples is to be found among polymethylene bis-trimethylammonium salts (Paton and Zaimis, 1949). Ganglion-blocking activity is maximal with hexamethonium (see formula 8,n = 6) while depolarising neuromuscular blocking activity is maximal with decamethonium (see formula 8, n = 10). Since acetylcholine is the natural neurotransmitter at both these sites and since its actions at both sites can be mimicked by, for example nicotine, it seems reasonable to assume that there is no very fundamental difference between the receptors at the neuromuscular junction and at the ganglion. If this assumption is correct then a possible explanation for differences in drug action at the two sites would seem to lie in different receptor distribution.

It is known that the blocking actions of hexamethonium and decamethonium are greater than those of the corresponding monoquaternary compounds. A reasonable inference to be drawn is that in both the ganglia and the neuromuscular junction are pairs of anionic sites capable of interacting with quaternary ammonium groups. The interquaternary distance of hexamethonium is 8.5 to 9 Å and of decamethonium, 14 Å. These distances may represent the separation of anionic sites in the ganglion and the neuromuscular junction respectively. The nature of these anionic sites is, of course, open to speculation. They may be relatively non-specific but it is tempting to argue that they form part of the acetylcholine receptor.

In an attempt to test this idea, Barrass, Brimblecombe, Rich and Taylor (1970) studied the actions of some acetylcholine homologues (see formula 50) where n = 3 to 12.

$$CH_3COO(CH_2)_n\overset{+}{N}(CH_3)_3$$

50

Their activities on the neuromuscular junction were studied using the isolated semi-spinalis muscle of the chick, a preparation described by Child and Zaimis (1960). Activities were also measured on the superior cervical ganglion of the anaesthetised cat using contraction of the nictitating membrane as the response.

On the isolated chick semi-spinalis all the compounds were stimulants, that is, they had a depolarising action. Peaks of activity were observed with the hexamethylene and dodecamethylene homologues, the latter being the most active of all the compounds studied. Compounds with three to ten methylene groups were all ganglion stimulants with a peak of activity at the hexamethylene homologue. Compounds with eleven and twelve methylene groups produced ganglion blockade at doses very near to those which stimulated the ganglion. These results are summarised in Table 3.1.

TABLE 3.1

The activities of acetylcholine homologues at the neuromuscular junction and on an autonomic ganglion

General formula $CH_3COO(CH_2)_n \overset{+}{N}(CH_3)_3$

(From Barrass *et al.* 1970)

n	Isolated chick semi-spinalis muscle. Potency relative to nicotine with 95% limits	Cat superior cervical ganglion. Approximate potency relative to acetylcholine
3	0.009 (0.008–0.01)	1
4	0.067 (0.062–0.072)	2
5	0.069 (0.057–0.077)	1
6	0.46 (0.43–0.49)	4
7	0.23 (0.22–0.24)	2.5
8	0.078 (0.066–0.087)	1
9	0.12 (0.11–0.13)	0.75
10	0.22 (0.20–0.25)	0.5
11	0.53 (0.48–0.57)	0.01*
12	0.84 (0.74–0.95)	0.01*

* produced ganglion blockade

There is a clear parallel here between the maximal agonist activity of the hexamethylene homologue on the ganglion and the dodecamethylene homologue at the neuromuscular junction and the respective maximal blocking activities at the same two sites of hexamethonium and decamethonium. It is possible that while the antagonist drugs interact with two receptors through their two quaternary groups, the agonist drugs do so by a quaternary group interacting with the anionic site of one receptor and an acetyl group with an appropriate receptor site on an adjacent receptor. This would not, however, account for the relatively high neuromuscular junction stimulant activity of the hexamethylene compound. A possibility is that the acetylcholine receptors are arranged in pairs in the ganglion whereas at the neuromuscular junction they are arranged in trios with two of their anionic sites separated by the interquaternary distance of hexamethonium and two by the interquaternary distance of decamethonium. Even this cannot explain all the facts, for hexamethonium would then be expected to show appreciable blocking activity at the neuromuscular junction.

In an attempt to obtain further relevant information, Barrass, Brimblecombe, Muir and Rich in unpublished work investigated other compounds comprising two identical units linked by a polymethylene chain of variable length. Hey in 1952 had shown that *m*-bromophenyl choline ether (see formula 51) had appreciable nicotinic activity; consequently a series of

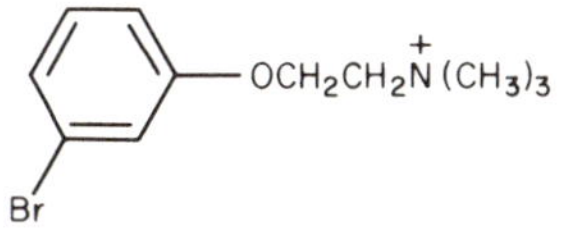

51

bis-quaternary *m*-bromophenyl choline ethers (see formula 52) with methylene chains varying in length from C_3 to C_{14} were studied for their actions at the neuromuscular junction and on autonomic ganglia.

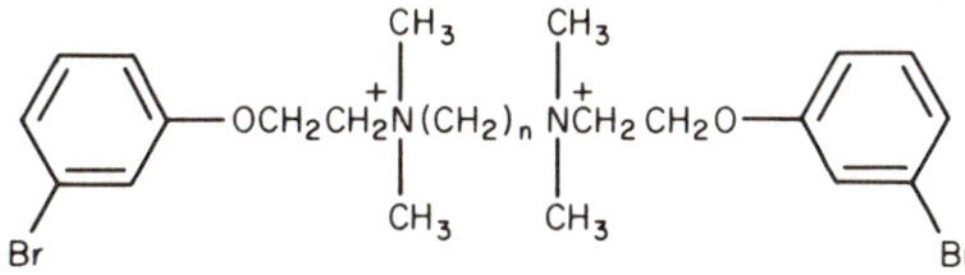

52

None of these bis-quaternary compounds showed stimulant activity at either the neuromuscular junction or on autonomic ganglia. In fact they acted as blocking agents at both sites although their actions on ganglia appeared to be non-specific. *In vivo*, peak neuromuscular blocking activity of the non-depolarising type was demonstrated in the cat tibialis preparations by the compounds with 6 to 9 atoms in the polymethylene chain. *In vitro*, the C_{12} compound showed the greatest activity in antagonising succinylcholine-induced contraction of the chick semi-spinalis muscle. These results do nothing to clarify the picture; obviously additional work is required.

In the meantime it should be pointed out that other authors (for example Barlow, 1964) have suggested that the distance between receptors at the neuromuscular junction makes it unlikely that a single drug molecule will interact with more than one at once. Waser (1959) calculated that there are about 10 receptors per end-plate in the mouse diaphragm corresponding to one receptor in an area of 2×10^4 Å^2. Such a distribution would not be compatible with an inter-receptor distance of 14 Å for example, as suggested above unless the receptors are grouped. That the latter may not be the case was suggested by the work of Miledi (1960).

At present, then, insufficient evidence exists to enable a firm opinion to be formed concerning similarities or dissimilarities between nicotinic receptors at the neuromuscular junction and at autonomic ganglia. For the present it is probably best to assume that they are essentially similar. Certainly in any discussion of possible central nicotinic receptors such as is found later in this monograph it does not seem feasible to attempt to distinguish different types of receptors.

3.4 The nature of nicotinic receptors

It seems that the structural requirements for nicotinic activity are not as stringent as those for muscarinic activity. This matter was discussed by Barrass *et al.* (1968) who studied the pharmacology of some amino-acid esters and amides (see formulae 47 and 53).

$$R_3\overset{+}{N}(CH_2)_nCONR'_2 . X^-$$

53

Several of the compounds had appreciable nicotinic activity and a comparison of molecular models of these with models of such nicotinic agents as acetylcholine, carbachol and the 4-oxopentyltrimethylammonium ion showed that in all cases a quaternary nitrogen atom was separated from a carbonyl oxygen atom by 4 to 4.5 Å. This suggests the presence of at least two active sites on the receptor separated by this distance. Later, Brimblecombe, Hunt, Rickard and Taylor (1969) studied some 1- and 2-substituted and 1,2-disubstituted, 1,4,5,6-tetrahydropyrimidines (see formula 54).

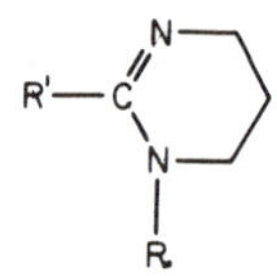

54

Several of these compounds possessed nicotinic activity, the most active being the 2-phenyl derivative (formula 54, R = H, R′ = Ph). With this compound the aromatic nucleus and the amidino residue in the pyrimidine ring would probably be capable of interaction with the two previously postulated receptor sites.

Later, in an attempt to produce compounds structurally closer to nicotine the Author and his colleagues prepared some cyclic amidines (see formula 55).

These again showed nicotinic activity and again it was considered that the aromatic nucleus and the amidino residue could be involved in a two-site drug-receptor interaction.

This model could well represent an over-simplification. There are some facts which are not very satisfactorily explained by it. For example, in the amino-acid esters and amides studied by Barrass *et al.* (1968) there were only very small differences in the nicotinic activities of corresponding

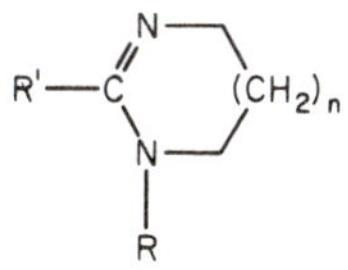

55

propionates and butyrates (see formula 47, n = 2 and 3) whereas ten-fold differences in activity were found between corresponding propionamides and butyramides (see formula 53, n = 2 and 3). Another difficulty lies in attempting to explain why nicotine and the tetrahydropyrimidines and other amidines referred to above are active nicotinic agents as tertiary bases while in the series of amino-acid esters and amides high nicotinic activity is only found with quaternary salts.

Earlier Hey (1952) studied a large number of compounds with nicotinic activity and reached the conclusion that maximum activity would be found in compounds as in formula 56 especially where a reduction in the electron density of the oxygen atom gave structures like 57.

$$RO.CH_2CH_2\overset{+}{N}(CH_3)_3 \qquad\qquad R^-:\overset{+}{O}CH_2CH_2\overset{+}{N}(CH_3)_3$$

56 57

Hey's results also seem to point to a two-site drug receptor interaction with the cationic head and an oxygen atom involved in compounds such as choline phenyl ether.

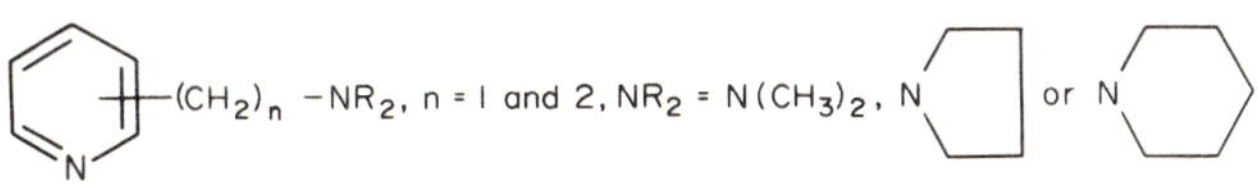

58

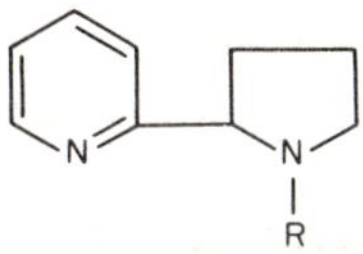

59

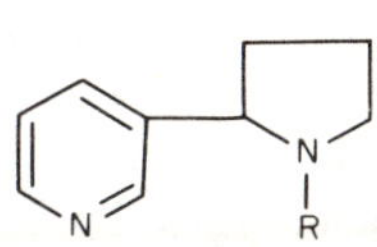

60

In another report Yamamoto (1965) compared the insecticidal activities of a series of pyridylpyrrolidines and pyridylpiperidines and found that substitution at the 3-position in the pyridine ring always resulted in higher activity than substitution in the 2-position. Haglid (1965) studied some N(pyridyl) methylpyrrolidones (see formula 61) and found the following gradation in potency: 3-pyridyl ≫ 4-pyridyl > 2-pyridyl.

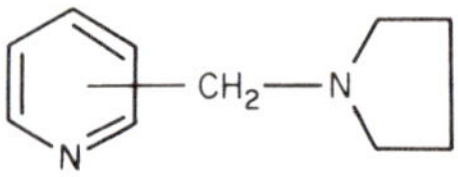

61

In the cyclic amidines (see formula 55) it was also found that attachment of a diazine ring (5,6 or 7 membered) at the 3 position of the pyridine ring resulted in compounds with the highest nicotinic activity, followed in turn by compounds where the attachement was at the 4 and 2 positions.

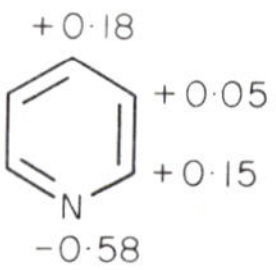

FIGURE 3.1 *Relative net charge difference in the atoms of the pyridine ring.*

It seems clear from all this evidence that two groups in the drug molecule are involved in interactions with the receptor but consideration must also be given to the fact that stereochemistry has some influence upon activity implying that a third group is involved. Pictet and Rotschy reported in 1904 that naturally occurring (−)-nicotine was about twice as toxic to guinea-pigs and rabbits as synthetic (+)-nicotine. In 1965 Domino reported a study of various effects of (−)- and (+)-nicotine and found that the latter was pharmacologically very weak being between four and seven times less active than the former in various tests. Hudson and Neuberger had already shown in 1950 that (−)-nicotine had the S-configuration.

Barlow (1965) suggested that in nicotine itself the third group which must be involved in drug-receptor interaction should be found in the part of the molecule which links the cationic head and the charged atom. That this third group is not essential for activity is shown by the fact that some analogues of nicotine are not optically active. Barlow considered, rather, that the third group might make the molecule inactive if in the wrong position.

The discussion so far has been based upon the assumption that a partial positive charge on one of the groups in the drug is necessary for nicotine activity. An alternative point of view was put forward by Sekul and Holland (1961) and Sekul, Holland and House (1963) following their studies of some esters of choline. They suggested that a partial negative charge at some distance from the onium group was essential for nicotinic activity. This negative charge would be situated in a position corresponding to that of the carbonyl oxygen atom of acetylcholine. Ormerod (1955) had previously studied a series of benzoylcholine esters and had observed that activity was enhanced by the incorporation of electron-releasing substituents into the molecule, a procedure which would, as is pointed out by Kier (1970), increase the partial negative charge on the carbonyl oxygen atom.

Triggle (1965) considered these apparently conflicting pieces of evidence and suggested that two secondary sites, one with a partial positive charge and one with a partial negative charge, might be necessary for activity. The oxygen atoms of acetylcholine would represent two such sites.

Molecular orbital theory would seem to constitute an ideal tool for studying this matter of the secondary charge and its relation to the onium group and Kier (1968) made such a study of the nicotinium ion using a method known as extended Huckel Theory. His calculations indicated two identically preferred conformations in which the pyridyl ring is perpendicular to the pyrrolidine ring. The pyridyl nitrogen is negatively charged. In these two conformations (see formula 62 a and b) the inter-nitrogen distances are 4.76 and 4.36 Å.

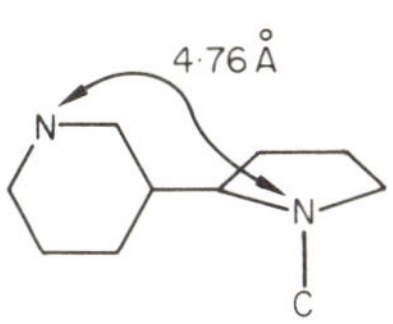

62a

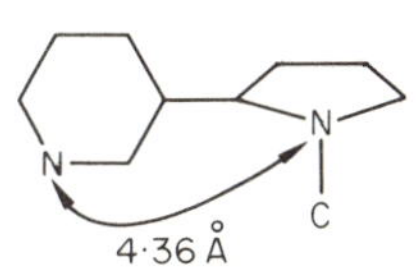

62b

Kier then compared these conformations with the preferred conformations of acetylcholine which he had calculated in 1967 and found that only one of them (62a) corresponded with any of the acetylcholine conformations. The corresponding acetylcholine conformation (see formula 63) was then considered to be the nicotinic conformation of that molecule.

The conclusions to be drawn from these results are that a partial negative charge is required on the so-called secondary binding site of the receptor and that it should be separated from the primary binding site by a distance of about 4.85 ±0.10 Å.

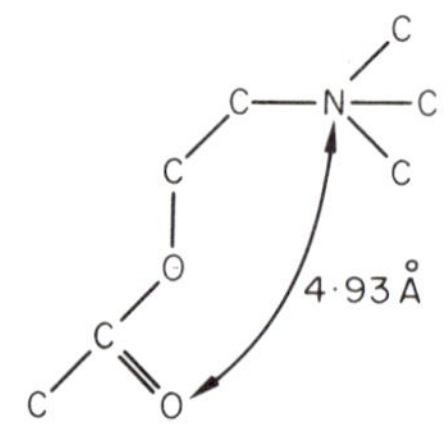

63

The matter still cannot be regarded as being completely settled however. Barlow (1964) points out an objection to the suggestion that a partial negative charge on the secondary binding site is of importance. The negative charge on the carbonyl oxygen atom will be affected by substituents in the esterifying acid more so than the partial positive charge on the carbonyl carbon atom, especially in a conjugated system. Yet activities in the benzoyl esters of choline are much less affected by substituents than are activities in the phenyl ethers.

In a subsequent chapter in this monograph drug effects at possible central nicotinic sites will be considered. In view of the current uncertainty as to possible differences between ganglionic and neuromuscular junction receptors and as to the nature of the receptor no attempt will be made to distinguish different types of receptors—they will all be designated "nicotinic".

References

Abdon, N. O. (1945). On the influence of atropine on nicotine-like actions of acetylcholine. *Acta. physiol. Scand.*, **1**, 153–170

Abood, L. G. (1968). The psychotomimetic glycollate esters. *Drugs affecting the Central Nervous System.* (ed. A. Burger) Marcel Dekker Inc., New York, pp 127–167

Abood, L. G. and Biel, J. H. (1962). Anticholinergic psychotomimetic agents. *Int. Rev. Neurobiol.*, **4**, 217–273

Barlow, R. B. (1964). *Introduction to Chemical Pharmacology*. Methuen & Co. Ltd., London

Barlow, R. B. (1965). Chemical structure and biological activity of nicotine and related compounds. *Tobacco Alkaloids and Related Compounds.* (ed. U. S. von Euler) Pergamon Press, Oxford, pp 277–301

Barlow, R. B. and Hamilton, J. T. (1962). Effects of some isomers and analogues of nicotine on junctional transmission. *Br. J. Pharmac. Chemother.*, **18**, 510–542

Barrass, B. C., Brimblecombe, R. W., Parkes, D. C. and Rich, P. (1968). The cholinergic properties of some amino-acid esters and amides. *Br. J. Pharmac.*, **34**, 245–257

Barrass, B. C., Brimblecombe, R. W., Rich, P. and Taylor, J. V. (1970). Pharmacology of some acetylcholine homologues. *Br. J. Pharmac.*, **39**, 40–48

Bentley, G. A. and Sabine, J. R. (1963). The effects of ganglion-blocking and postganglionic sympatholytic drugs on preparations of the guinea-pig vas deferens. *Br. J. Pharmac. Chemother.*, **21**, 190–201

Beranek, R. and Vyskocil, F. (1967). The action of tubocurarine and atropine on the normal and denervated rat diaphragm. *J. Physiol. Lond.*, **188**, 53–66

Beranek, R. and Vyskocil, F. (1968). The effects of atropine on the frog sartorius neuromuscular junction. *J. Physiol. Lond.*, **195**, 493–503

Botkin, S. (1862). Uber die physiologische Wirkung des schwefelsauren Atropins. *Virchows Arch. path. Anata. Physiol.*, **24**, 83–92

Brimblecombe, R. W. and Everett, S. D. (1969). Actions of a cholinergic antagonist on mammalian skeletal muscle. *Br. J. Pharmac.*, **36**, 172–173P

Brimblecombe, R. W. and Everett, S. D. (1970). Actions of some cholinergic antagonists on fast-twitch and slow-twitch skeletal muscle of the cat. *Br. J. Pharmac.*, **40**, 45–56

Brimblecombe, R. W. and Green, D. M. (1968). The peripheral and central actions of some anticholinergic substances. *Int J. Neuropharmac.*, 7, 15–21

Brimblecombe, R. W. and Sutton, J. V. (1968). The ganglion-stimulating effects of some amino-acid esters. *Br. J. Pharmac.*, **34**, 358–369

Brimblecombe, R. W. and Webb, S. N. (1973). Further studies on the interaction between acetylcholine antagonists and anticholinesterases on cat soleus muscle. *Br. J. Pharmac.*, **48**, 361–362

Brimblecombe, R. W., Green, D. M., Stratton, J. A. and Thompson, P. B. J. (1970). The protective actions of some anticholinergic drugs in sarin poisoning. *Br. J. Pharmac.*, **39**, 822–830

Brimblecombe, R. W., Hunt, R. R., Rickard, R. L. and Taylor, J. V. (1969). The synthesis and pharmacology of some 1,4,5,6-tetrahydropyrimidines. *Br. J. Pharmac.*, **37**, 425–435

Brown, D. A. (1968). Effects of pre-ganglionic denervation on the sensitivity of sympathetic ganglion cells to stimulant compounds. *J. Physiol. Lond.*, **196**, 117–118P

Brown, D. A., Halliwell, J. V., Jones, K. B. and Quilliam, J. P. (1970). An attempt to discover whether acetylcholine can release acetylcholine from a sympathetic ganglion. *Br. J. Pharmac.,* **38**, 446–448P

Bulbring, E. (1946). Observations on the isolated phrenic nerve-diaphragm preparation of the rat. *Br. J. Pharmac. Chemother.*, **1**, 38–61

Buttle, G. A. H. and Zaimis, E. J. (1949). The action of decamethonium iodide in birds. *J. Pharm. Pharmac.*, **1**, 991–992

Child, K. J. and Zaimis, E. J. (1960). A new biological method for the assay of depolarising substances using the isolated semi-spinalis muscle of the chick. *Br. J. Pharmac. Chemother.*, **15**, 412–416

Cooper, G. H., Green, D. M., Rickard, R. L. and Thompson, P. B. (1971). The ganglion-blocking activity of diastereoisomeric dimethylamino-bornyl acetates and their methiodides. *J. Pharm. Pharmac.*, **23**, 662–670

Cushny, A. R. (1903). Atropine and the hyoscyamines—a study of the action of optical isomers. *J. Physiol. Lond.,* **30**, 176–194

Dale, H. H. (1914). The actions of certain esters and ethers of choline and their relation to nicotine. *J. Pharmac. exp. Ther.,* **6**, 147–190

Dale, H. H. and Gaddum, J. H. (1930). Reactions of denervated voluntary muscle and their bearing on the mode of action of parasympathetic and related nerves. *J. Physiol. Lond.,* **70**, 109–144

Domino, E. F. (1965). Some comparative pharmacological actions of (–) –nicotine, its optical isomer, and related compounds. *Tobacco Alkaloids and Related Compounds.* (ed. U. S. von Euler) Pergamon Press, Oxford, pp 303–313

Dutta, N. K. (1949). Some pharmacological properties common to atropine, pethidine, procaine and quinidine. *Br. J. Pharmac. Chemother.*, **4**, 197–201

Erdtman, H., Haglid, F. and Wellings, I. (1963). Synthetic analogues of nicotine. *Acta. Chem. Scand.,* **17**, 1717–1726

Falk, G. (1961). Electrical activity of skeletal muscle—its relation to the active state. *Biophysics of Physiological and Pharmacological Actions.* (ed. A. M. Shanes) American Association for the Advancement of Science, publication No. 69, Washington, D.C., pp 259–279

Haffner, F. (1914). Uber die Wirkung des atropins auf den Skelettmuskel. *Arch. int. Pharmacodyn. Ther.*, **24**, 547–554

Haglid, F. (1965). Synthetic analogues of nicotine. *Tobacco Alkaloids and Related Compounds.* (ed. U. S. von Euler) Pergamon Press, Oxford, pp 315–319

Harvey, A. M. (1939). The actions of procaine on neuromuscular transmission. *Bull. Johns Hopkins Hosp.,* **65**, 223–238

Hey, P. (1952). On relationship between structure and nicotine-like stimulant activity in choline esters and ethers. *Br. J. Pharmac. Chemother.*, 7, 117–129

Hucović, S. (1961). Responses of the isolated sympathetic nerve-ductus preparation of the guinea pig. *Br. J. Pharmac. Chemother.*, **16**, 188–194

Hudson, C. S. and Neuberger, A. (1950). The stereochemical formulas of the hydroxyproline and allohydroxyproline enantiomorphs and some related substances. *J. org. Chem.*, **15**, 24–34

Jaco, N. T. and Wood, D. R. (1944). The interaction between procaine, cocaine, adrenaline and prostigmine on skeletal muscle. *J. Pharm. exp. Ther.*, **82**, 63–73

Kier, L. B. (1967). Molecular orbital calculation of preferred conformations of acetylcholine, muscarine and muscarone. *Mol. Pharmac.*, **3**, 487–494

Kier, L. B. (1968). A molecular orbital calculation of the preferred conformation of nicotine. *Mol. Pharmac.*, **4**, 70–76

Kier, L. B. (1970). Receptor mapping using molecular orbital theory. *Fundamental Concepts in Drug Receptor Interactions*. (eds. J. F. Danielli, J. F. Moran and D. J. Triggle) Academic Press, New York, pp 15–45

Koelle, G. B. (1970). Neuromuscular blocking agents. *The Pharmacological Basis of Therapeutics*. 4th edition (eds. L. S. Goodman and A. Gilman) Macmillan, New York, pp 601–619

Koelle, W. A. and Koelle, G. B. (1959). The localisation of external or functional acetylcholinesterase at the synapses of autonomic ganglia. *J. Pharmac. exp. Ther.*, **126**, 1–8

Lammers, W. and Ritchie, J. M. (1955). The action of quinine and quinidine on the contractions of striated muscle. *J. Physiol. Lond.*, **129**, 412–423

Longuet-Higgins, H. C. and Coulson, C. A. (1947). A theoretical investigation of the distribution of electrons in some heterocyclic molecules containing nitrogen. *Trans. Farad. Soc.*, **43**, 87–94

Miledi, R. (1960). Junctional and extra-junctional acetylcholine receptors in skeletal muscle fibres. *J. Physiol. Lond.*, **151**, 24–30

Ormerod, W. E. (1956). The pharmacology of benzoylcholine derivatives and the nature of carbonyl receptors. *Br. J. Pharmac. Chemother.*, **11**, 267–272

Paton, W. D. M. and Zaimis, E. J. (1949). The pharmacological actions of polymethylene bistrimethylammonium salts. *Br. J. Pharmac. Chemother.*, **4**, 381–400

Pictet, A. and Rotschy, A. (1904). Synthèse des nicotins. *Ber. deut. Chem. Ges.*, **37**, 1125–1235

Segawa, T., Kojima, M. and Takagi, H. (1967). Effect of several anticholinergic drugs on the post-tetanic potentiation in the phrenic nerve-diaphragm preparation of the rat. *Jap. J. Pharmac.*, **17**, 465–474

Sekul, A. A. and Holland, W. C. (1961). Comparative pressor effect of certain unsaturated acid esters of choline. *J. Pharmac. exp. Ther.*, **133**, 313–318

Sekul, A. A., Holland, W. C. and House, R. (1963). Relationship between electronic structure and intensity of nicotine-like action of choline esters. *Arch. int. Pharmacodyn. Ther.*, **141**, 404–411

Triggle, D. J. (1965). *Chemical Aspects of the Autonomic Nervous System.* Academic Press, New York

Volle, R. L. (1966). Muscarinic and nicotinic stimulant actions of autonomic ganglia. *International Encyclopedia of Pharmacology and Therapeutics. Section 12. Ganglionic Blocking and Stimulating Agents.* (ed. A. H. Karczmar) Pergamon Press, Oxford

Volle, R. L. and Koelle, G. B. (1961), The physiological role of acetylcholinesterase (AChE) in sympathetic ganglia. *J. Pharmac. exp. Ther.*, **133,** 223–240

Volle, R. L. and Koelle, G. B. (1970). Ganglionic stimulating and blocking agents. *The Pharmacological Basis of Therapeutics.* 4th Edition. (eds. L. S. Goodman and A. Gilman) Macmillan, New York, pp 585–600

Waser, P. G. (1959). *Curare and Curare-like Agents.* (ed. D. Bovet) Elsevier, Amsterdam, pp 244–264

Yamamoto, I. (1965). Nicitinoids as insecticides. *Adv. in Pest. Control Res.*, **6**, 231–260

4 Anticholinestarase Agents

In this chapter, unlike the two previous ones, both the peripheral and central actions of anticholinesterase agents will be discussed. These agents have attracted a great deal of attention, certainly out of all proportion to their medicinal use, which is limited. They have, however, proved to be very valuable tools in studies of the processes involved in cholinergic transmission. Therapeutically there are only three types of conditions for which these agents are generally accepted as being useful (Koelle, 1970). These are atony of the smooth muscle of the intestinal tract and bladder, myasthenia gravis and glaucoma. Neostigmine (see formula 65) appears to be the drug of choice for the first two conditions while for the latter both carbamates and organophosphorus compounds have been used.

For measuring the potency of anticholinesterase agents *in vivo* various overt manifestations of cholinergic overactivity, such as miosis or salivation, have been used as the end-points in assays. More usually, and in most circumstances more usefully, the *in vitro* activity is measured and is often expressed as a pI50 value, that is the negative logarithm of the molar concentration of the inhibitor required to give 50 % inhibition of the enzyme. However, this does not give complete information about the type of inhibition and a more definitive parameter is k_2, the bimolecular rate constant, expressed in $1\ min^{-1}\ mol^{-1}$. For more detailed discussion of these matters reference should be made to Barlow (1964) and Usdin (1970). The *in vitro* methods usually depend upon the measurement of liberated hydrogen ions as the hydrolysis of acetylcholine is catalysed by acetylcholinesterase.

Anticholinesterase agents, then, may be defined as inhibitors of cholinesterases. Attention in this chapter will be focused on those agents which specifically inhibit acetylcholinesterase and it is conventional to classify the inhibition as being either reversible or irreversible in nature. These terms have led to some confusion depending on the way in which they are defined. Some workers have defined reversibility essentially in terms of the ease with which the inhibited enzyme can be reactivated, or, put another way, the ease with which the inhibitor can be removed. Used in this way the terms reversible and irreversible cannot be considered as being absolute. For example, the carbamates have often been classified as reversible inhibitors for, as they will be indicated below, they form a

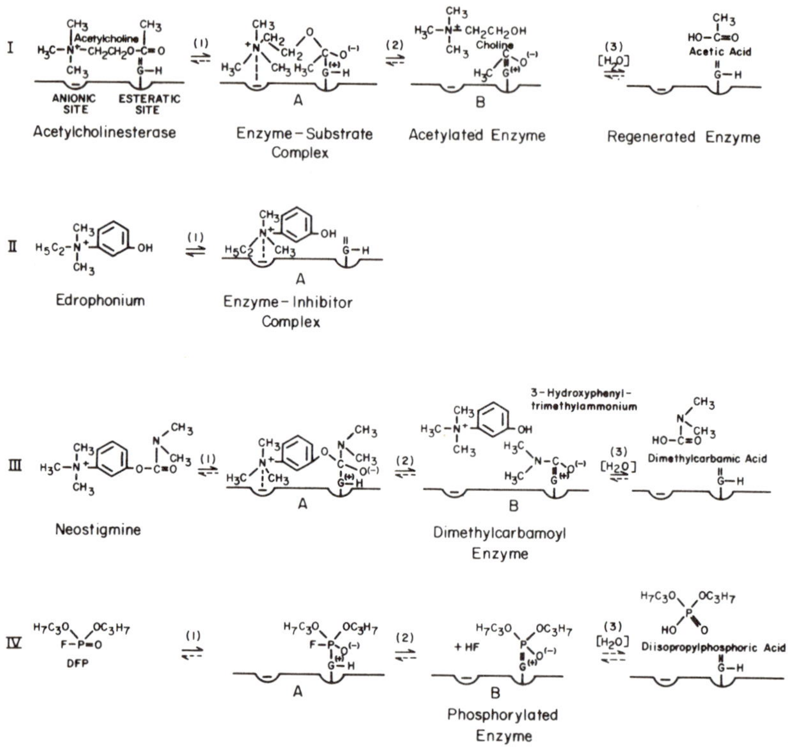

FIGURE 4.1 *Steps involved in the hydrolysis of acetylcholine (ACh) by acetylcholinesterase (AChE) (i), and in the inhibition of AChE by reversible (ii), carbamyl ester (iii), and organophosphorus (iv) agents.*

(i) The substrate, ACh, combines with an active unit of the enzyme (step 1) to form a complex (A) by electrostatic attraction between the N atom of the choline moiety and the anionic site of the enzyme, and the electrophilic C atom of the carboxyl group and a protonated acidic group (–G–H) of the esteratic site; choline is then split off (step 2) leaving the acetylated enzyme (B) which reacts rapidly with water (step 3) to produce acetic acid and the regenerated active enzyme.

(ii) Edrophonium combines only at the anionic site (step 1) to form a rapidly reversible enzyme-inhibitor complex (A).

(iii) Neostigmine and related ammonium- or amino-carbamyl esters react with the enzyme in the same manner as does the substrate (steps 1 and 2); however, the dimethylcarbamoyl enzyme (III–B) reacts with water (step 3) at less than a millionth the rate of the corresponding acetylated form (I–B)

(iv) DFP and similar organophosphorus inhibitors react only at the esteratic site to form a phosphorylated enzyme; in the case of diisopropylphosphoryl-AChE (IV–B), essentially no spontaneous hydrolytic reactivation occurs.

Heavy, light, and dash arrows represent extremely rapid, intermediate, and extremely slow or insignificant reaction velocities, respectively.

Goodman and Gilman, 1970. (Reprinted by permission of The Macmillan Co., New York.)

carbamylated enzyme which is relatively susceptible to hydrolytic reactivation. However, this is not reversible inhibition as defined by, for example, Usdin (1970) who states that "the ability to regenerate the free enzyme with decomposition of the inhibitor is probably one of the most characteristic features of irreversible inhibition". The point is that the reactivation of the carbamate-inhibited enzyme involves the release of carbamic acid rather than of the unchanged inhibitor. Thus, it seems reasonable, according to Usdin, to regard organophosphorus compounds, carbamates and organosulphates as irreversible inhibitors leaving a miscellany of compounds, including oximes, certain inorganic compounds, some bis- and mono-quaternary compounds and a miscellaneous group, as reversible inhibitors. Usdin gives a table listing many of these compounds together with their enzyme-inhibiting activities. Comprehensive lists of organo-phosphorus compounds are also given by Holmstedt (1959, 1963).

The steps involved in the hydrolysis of acetylcholine by acetylcholinesterase and in the inhibition of the enzyme by edrophonium (see formula 64), (a reversible inhibitor), neostigmine (see formula 65), and DFP or dyflos (see formula 66) (irreversible inhibitors) are summarised in Figure 4.1.

Points to be noted particularly from this Figure are the different rates at which hydrolytic reactivation of the acetylated, carbamylated and phosphorylated enzymes occur. Despite the fact that reactivation of the carbamylated enzyme occurs at less than a millionth the rate of the acetylated enzyme, nevertheless it is sufficiently rapid for there to be no requirement for chemical reactivation to form part of the therapeutic procedure for poisoning with carbamates. In the case of the organophosphorus inhibitors, on the other hand, an essential part of the treatment is chemical reactivation of the inhabited enzyme. This is discussed in more detail below.

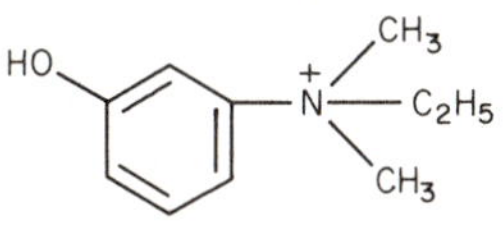

64

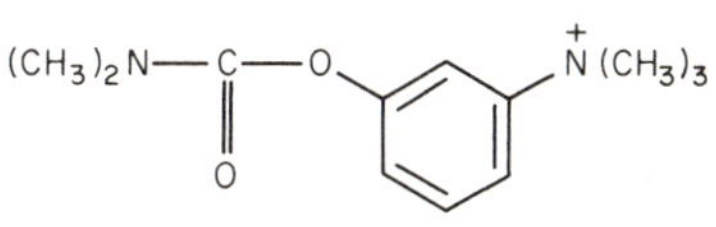

65

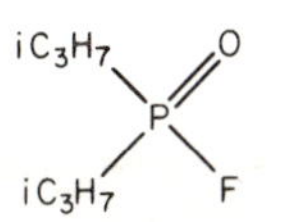

66

4.1 Acute pharmacological effects of anticholinesterases

Without any doubt the main acute effects of all anticholinesterase agents can be attributed to the accumulation of acetylcholine at sites where it functions as the chemical transmitter, that is, at postganglionic parasympathetic endings (muscarinic sites), at autonomic ganglia and the skeletal neuromuscular junction (nicotinic sites), and in the central nervous system (central sites). It is possible to relate all the major signs and symptoms of poisoning with anticholinesterase agents to actions at these sites. The following represents a fairly comprehensive list but it must be realised that the occurrence of signs and symptoms depends to a great extent on the dose of agent, the route by which it has been absorbed and on the nature of the agent itself. Nevertheless, in general, poisoning with a variety of inhibitors has been reported as producing essentially the same clinical picture (Wills, 1970).

4.1.1 Muscarinic actions

Miosis is one of the earliest signs of poisoning when the anticholinesterase agent has direct access to the eye in the form of a vapour or aerosol. After systemic administration the effect occurs later and may not be so significant. There is a generalised stimulation of smooth muscle resulting in bronchoconstriction, contractions and increased peristalsis in the stomach and intestine and contraction of the muscles in the walls of the genito-urinary tract. There has been some dispute (Cullumbine, 1963) as to whether the bronchoconstriction is due primarily to accumulation of acetylcholine at the vagal ganglion or at the post-ganglionic endings. There is a generalised stimulation of secretion by lachrimal, sweat, salivary, bronchial, gastric and pancreatic glands all of which have a cholinergic innervation. The excess secretions of the salivary and bronchial glands, in particular, can lead to partial blockage of the airway and respiratory embarrassment. Effects on the cardiovascular system are manifested in general by a slowing of the heart and a fall in blood pressure but there have been many variations reported on this basic response due, undoubtedly, to the complex of reflexes and feed-back mechanisms which operate to control heart rate and blood pressure. The bradycardia is due, in part, to the accumulation of acetylcholine at the vagal endings, that is, a muscarinic action, but there also seems to be some contribution to the effect from a parasympathetic ganglion stimulation—a nicotinic effect. Evidence concerning the mechanism whereby the hypotension is produced is somewhat conflicting (Cullumbine, 1963). Two factors seem to be operating, a decreased peripheral resistance due to vasodilatation and a reduced cardiac output to the bradycardia. The relative importance of these two factors varies with the species being studied and also with the agent being used.

4.1.2 Nicotinic actions

Skeletal muscles and autonomic ganglia are stimulated by low doses of anticholinesterase agents and depressed by higher doses. Because of these dose-dependence effects, it is difficult to give a generally applicable composite picture. Difficulties in accommodation occur since the ciliary muscle contracts making the lens relatively spherical and thus accommodated for near vision. There are also fasiculations of the muscles of the eyelid and these two effects added to the miosis, make for considerable difficulties in vision. The complex actions of anticholinesterases at autonomic ganglia were reviewed in detail by Zaimis (1963) who concluded that the pharmacological actions of these drugs were due to accumulation of acetylcholine at the ganglia but that the resultant effects were less prominent and less easy to demonstrate than those at the neuromuscular junction. It is still not clear, for example, as to whether the hypertension which often follows the hypotension referred to above (or replaces it in some species) is due to over-activity of the sympathetic ganglia or to a more direct vasoconstriction action. In general terms the effects at the neuromuscular junction are easily described; muscular fasciculations are followed by muscular weakness and eventual neuromuscular blockade. The detailed actions which result in these effects are, however, complex and still not completely understood. The basic difficulty lies in assessing the relative contributions of inhibition of cholinesterase and direct actions of the drugs to the effects observed. This matter was reviewed fully by Werner and Kuperman (1963).

4.1.3 Central actions

In general the central actions of anticholinesterases can be explained in terms of stimulation followed by inhibition. Thus, stimulant effects like apprehension, dizziness, tremor, ataxia and convulsions are succeeded by, or accompanied by, inhibitory effects which culminate in coma. Changes in the electroencephalogram (EEG) seem characteristic with a desynchronised pattern of activity with a low amplitude, high frequency wave form giving way to a pattern with higher amplitude which in turn is followed by spike discharge or convulsive activity. Effects on the respiratory centre also, are characteristic with respiratory stimulation being succeeded by respiratory depression and eventual respiratory failure. The behavioural consequences of anticholinesterase poisoning will be discussed in more detail later in this chapter.

The cause of death is respiratory failure due in part to muscarinic actions (broncho-constriction and excess secretions into the airway), nicotinic actions (neuromuscular blockade of the respiratory muscles) and central actions of failure of the respiratory centre). Therapeutic or protective procedures, which are detailed below, are aimed, in the main, at reversing or preventing these actions.

4.1.4 Effects resulting from actions other than the inhibition of cholinesterase
Several references have been made earlier in this chapter to possible direct actions of anticholinesterase agents and, while there is no need to modify the statement that their main effects result from enzyme inhibition, a brief mention of these possible direct actions is appropriate here.

The best documented evidence relates to actions at the neuromuscular junction and especially to the ability of anticholinesterase agents to reverse the neuromuscular block produced by curare. Blaschko, Bulbring and Chou (1949) compared the anticholinesterase activities of a series of carbamates with their anticurare activities on the rat phrenic nerve-diaphragm preparation and were able to show a statistically significant correlation between these two sets of activities. This finding seemed to indicate that the reversal of curare block was due entirely to the anticholinesterase actions of the compounds. However, a few years later the data of Blaschko *et al.* were re-examined by Riker (1953) who pointed out that if the activities of individual compounds were compared there were certain clear discrepancies between anticurare activity and anticholinesterase activity which were masked by the procedure of comparing the activities of all twelve compounds which were studied. This was suggestive of a direct action of the carbamates and evidence has gradually accumulated to make it virtually certain that such actions exist. For example, Thesleff (1960) was able to show that neostigmine and physostigmine (see formula 67) in relatively high concentrations produced a neuromuscular blockade in the rat phrenic nerve-diaphragm preparation. This blockade occurred even if the nerve-muscle preparation had been treated with botulinum toxin which results in decrease of acetylcholine output. Thus it seems unlikely that the neuromuscular blockade by these two inhibitors can be explained entirely in terms of acetylcholine accumulation. If this were the case botulinum toxin pretreatment would be expected to reduce susceptibility to neuromuscular block.

A rather similar finding had been reported earlier by Riker and Wescoe (1946). They found that during *in vivo* experiments neostigmine was capable of causing a contraction of muscle even if this had been virtually depleted of acetylcholinesterase by chronic denervation or by pretreatment with dyflos. These authors found a difference between neostigmine and phystigmine in that the latter drug did not cause muscle contraction under these circumstances.

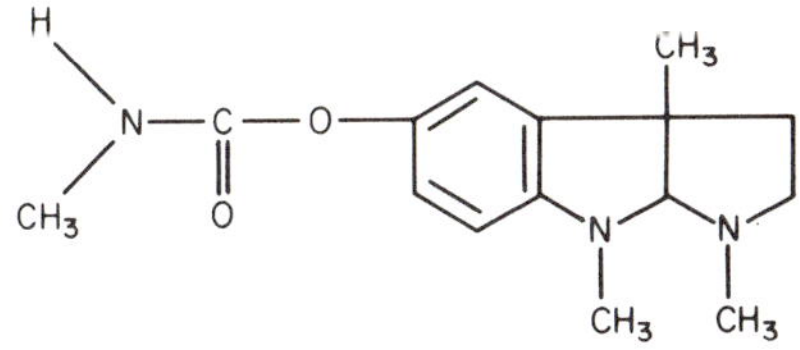

67

There is also a good deal of evidence, reviewed in detail by Werner and Kuperman (1963) to show that organophosphorus compounds, as well as carbamates, have actions at the neuromuscular junction not related to their anticholinesterase activity. One example will suffice. Barstad (1956) reasoned that a neuromuscular block due solely to accumulation of acetylcholine should be capable of reversal by curare. In practice he found using dyflos that neuromuscular transmission in the phrenic nerve-diaphragm preparation could only be restored by tubocurarine after moderate doses of the inhibitor. After higher doses tubocurarine only enhanced the block but after removal of the dyflos neuromuscular transmission was easily restored.

4.2 Chronic effects of anticholinesterases

Certain organophosphorus anticholinesterase agents give rise to delayed neurotoxic effects in man. These are characterised by a polyneuritis, with weakness of the musculature of the legs and arms progressing to flaccid paralysis of these extremities and, in some cases, leading eventually to a spastic paralysis. The condition is accompanied by degeneration of the myelin sheaths and axons of the spinal cord, somatomotor nerves and medulla. The effects are usually delayed in onset and either fatal, permanent or of very considerable duration. The classic example of an agent responsible for these effects is triorthocresyl phosphate (TOCP, see formula 68) and it has been estimated (Wills, 1970) that at least 35,000 people in various parts of the world have been paralysed to a lesser or greater degree by this substance. One of the largest outbreaks occurred in the Southern United States of America in 1930 during the prohibition era. During this period Jamaican ginger was in great demand as a beverage. Some preparations were adulterated with

TOCP and gave rise to the condition which became known as "Ginger Jake paralysis".

$$(CH_3O)_3\,P{=}O$$

68

Detailed accounts of the neurotoxicity of organophosphorus compounds have been written by Davies (1963); Casida, Baron, Eto and Engel (1963); Aldridge and Barnes (1961, 1966a, b), and no attempt will be made here to review the whole field, especially as it will be made apparent that this action of the compounds appears not to be related to their cholinesterase inhibiting activities; they should not therefore figure largely in a monograph dealing with drug actions on cholinergic systems.

One of the problems in investigating these neurotoxic effects is that marked differences exist between species and only the chicken and the cat seem to present a clinical picture in any way analogous to that seen in man. Most experimental work has been carried out with chickens and, as a result, it has been concluded (Davies, 1963) that two types of compounds are neurotoxic. These are certain triarylphosphates and certain alkyl organophosphorus compounds. The former compounds produce virtually no acute effects at the doses which cause paralysis while in general, the latter only produce neurotoxic effects at doses which are very near the lethal dose and which give rise to severe acute symptoms. In practice paralysis can only be produced by the alkyl compounds in animals which have been protected from acute lethal effects by the standard treatment of atropine and oxime (see below). Apparent uneventful recovery is then followed a few days later by the development of ataxia and paralysis. Alternatively the paralysis can be induced by the repeated administration of sublethal doses.

One of the early theories as to the mechanism of the production of neurotoxicity emanated from the work of Earl and Thompson (1952 a, b). It was suggested that inhibition of butyrylcholinesterase was responsible for the effect since the three known neurotoxic agents at that time, TOCP (see formula 68), dyflos (see formula 66) and Mipafox (see formula 69) were all relatively specific inhibitors of this enzyme.

However, later work showed that other specific inhibitors of butyryl-

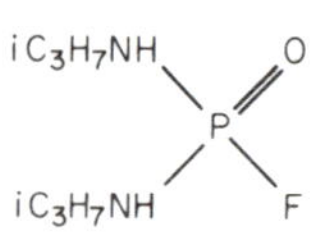

69

cholinesterase did not produce neurotoxicity and it has been concluded (Davies, 1963) that neither prolonged nor short-term inhibition of this enzyme in the central nervous system is necessarily a preliminary to the production of paralysis.

Another possible role of cholinesterase was suggested by Davies, Holland and Rumens (1960). They suggested that the enzyme effected the liberation of fluorine from alkyl organophosphorus compounds and that the former was in some way responsible for the biochemical lesion. The validity of this suggestion is however, open to considerable doubt since it in no way explains the neurotoxicity of the triarylphosphates which, according to Cavanagh (1964) produce effects in the cat and the hen identical to those of alkyl organophosphorus compounds.

The concensus of opinion then supports the view that demyelination and paralysis do not result from cholinesterase inhibition. The mechanism which is involved has still not been elucidated. Aldridge, Barnes and Johnson (1969) have considered the possibility that a protein which can be phosphorylated by neurotoxic agents is involved in the genesis of neurotoxicity. Cavanagh (1964) suggested that interference with the transference or storage of energy within nerve cells gives rise to the demyelination. Either or both of these suggestions may prove to be correct but much more experimental evidence is required before they can gain general acceptance.

4.3 Treatment of anticholinesterase poisoning

Occasionally overdosage with anticholinesterase agents occurs during their medicinal use for the conditions mentioned above. Much more commonly, accidental over-exposures to anticholinesterase insecticides have occurred and the potential use of organophosphorus agents in chemical warfare as the "nerve-agents" could constitute a considerable therapeutic or prophylactic problem. Consequently a considerable amount of work has been carried out on the treatment of poisoning with these agents.

If considerations of devices like protective clothing and respirators and external decontaminating agents are left aside as being outside the scope of this book then treatment must be directed at either protecting the acetylcholinesterase from the inhibitor, releasing an active enzyme from the enzyme-inhibitor complex or minimising the actions of the accumulated acetylcholine which is responsible for all the main effects of anticholinesterases. Theoretically the following types of procedure might be expected to be useful.

(1) Internal decontamination of the inhibitor before it can inhibit significant proportions of the enzyme

(2) Protection of the enzyme from the inhibitor
(3) Reactivation of the enzyme
(4) Inhibition of the release of acetylcholine or acceleration of its destruction
(5) Antagonism of the effects of acetylcholine
(6) Symptomatic treatment and miscellaneous drugs.

Of these possibilities, (1) and (4) have proved to be of little or no value in practice and so will not be discussed here. Excellent accounts of various aspects of the treatment of poisoning by anticholinesterases are given by Wills (1963, 1970), Hobbiger (1963), Grob (1963) and Koelle (1970) and all these have been drawn upon to some extent in the following account.

It should be remembered that the main problems arise with the organophosphorus inhibitors where the effects are long (possibly days or even weeks) in comparison with the carbamates where the effects are unlikely to persist for more than a few hours because of the relatively rapid rate at which spontaneous reactivation of the enzyme occurs. One other fact should be borne in mind, that maintenance of a patent airway and artificial ventilation are most useful adjuncts to, or if drugs are not available, substitutes for, drug treatment. With the carbamates particularly, adequate ventilation can protect against the effects of several lethal doses.

4.3.1 Protection of the enzyme from the inhibitor

The only feasible way of protecting acetylcholinesterase from an irreversible inhibitor is to use another compound which will react with the enzyme. This will in itself inhibit the enzyme so it is necessary to ensure that the inhibition is of a readily reversible type. Truly reversible inhibitors of the edrophonium type are not likely to be of much practical value since the length of time that they will protect the enzyme from irreversible inhibition will be very short indeed. The carbamates, on the other hand, with a duration of action intermediate between that of the true reversible inhibitors and that of the organophosphorus compound might be expected to inhibit some of the acetylcholinesterase in vital organs and tissues which are known from the work of Barstad, (1960) and Van der Meer and Wolthius (1965) to contain more enzyme than is necessary for normal functioning. This inhibited enzyme would then not be available for inhibition by an organophosphorus inhibitor but would be spontaneously reactivated and, providing the organophosphorus compound were not present in large excess, might be expected to be sufficient to maintain life, especially if used in conjunction with other therapeutic drugs. It must be assumed that the carbamate to be effective needs to be given before poisoning with the organophosphorus compound.

Koster first demonstrated that this theoretical possibility was a practical proposition in 1946. He showed that cats pretreated with physostigime (see formula 67) were resistant to the lethal effects of several LD50s of dyflos. Callaway (quoted by Berry and Davies, 1970) showed a similar protective action in rats and Barnes (quoted by Berry and Davies, 1970) also using rats, found that either neostigmine (see formula 65) or 1,5-di(4-ethyldimethylammoniumphenyl) pertan-3-one-diioxide (see formula 70) were as effective as physostigmine.

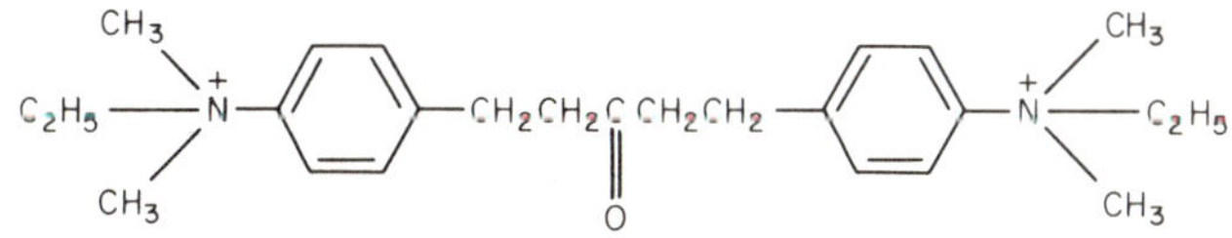

70

Koelle (1946) seems to indicate that the main contributory factor to the protective action of physostigmine or neostigmine is their ability to protect the enzyme from the organophosphorus inhibitor. He showed that *in vitro* both of these carbamates could protect acetylcholinesterase from inhibition by dyflos. Later, during *in vivo* experiments in rats, Dubois, Doull, Salerno and Coon (1949) were able to show that physostigmine pretreatment decreased the degree of enzyme inhibition by p-nitrophenyl diethyl thiophosphate (parathion; see formula 71) in both the peripheral and central nervous systems.

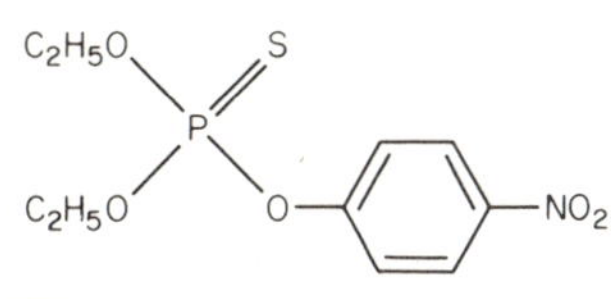

71

In these experiments an intraperitoneal dose of 5 mg/kg parathion produced a mean inhibition of brain cholinesterase activity of 85 per cent and of submaxillary gland enzyme activity of 71 per cent. A dose of 5 mg/kg physostigmine caused corresponding inhibitions of 12 and 49 per cent respectively. Animals which were given physostigmine 30 minutes before parathion showed inhibitions of 37 and 45 per cent respectively.

Despite experimental findings of this kind a number of workers have produced evidence which seems to indicate that the protective actions of carbamates against the lethal effects of organophosphorus compounds cannot be explained solely in terms of protection of the enzyme. Much of the evidence is not particularly convincing but that of Scaife (1960) seems to warrant serious consideration. He compared the *in vitro* acetylcholinesterase inhibiting activities of a number of compounds with their *in vitro* activities in protecting the enzyme from inhibition by ethyl-N,N-dimethyl phosphoramidocyanidate (tabun; see formula 72) or diethyl-S-diethylaminoethyl phosphorothiolate (DSDP, amiton; see formula 73).

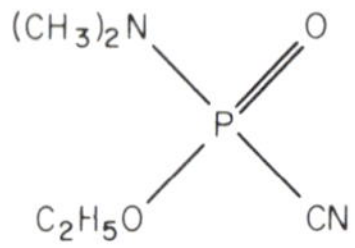

72

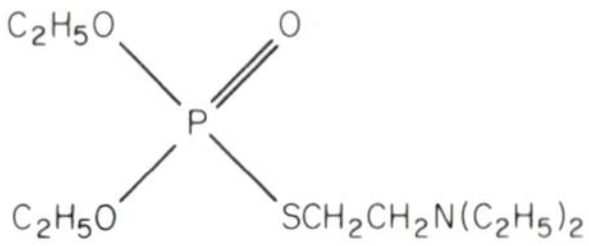

73

From his results Scaife concluded that the protection of acetylcholinesterase afforded by these compounds was not related to their affinity for the enzyme.

Various quaternary ammonium compounds have also been shown to confer a certain degree of protection against the lethal effects of anticholinesterase agents and while it is probable that this protection results, in the main, from their curare-like actions, many of them also possess some enzyme inhibiting activity which may be of some significance. Indeed, Despierre and Martin (1958) studied the ability of a number of quaternary ammonium compounds to protect human erythrocyte acetylcholinesterase from inhibition be neostigmine. Among these compounds there was a general, although not precise, parallelism between protective action and ability to inhibit the enzyme. In passing it is also worth noting that some authors (for example, Van Meter and Karczmar, 1968) have attributed at least part of the prophylatic action of benzoquinonium and oxamides, which will be discussed in more detail later, to their reversible inhibition of the enzyme.

There are, as will become apparent in the next section of this chapter, some organophosphorus inhibitors which form a complex with the enzyme which is not susceptible to oxime reactivation. Clearly, treatment of poisoning with such agents presents a considerable problem and any prophylactic procedure which might protect the enzyme from inhibition would be of great value. The nerve agent pinacolyl-methylphosphorofluoridate (soman; see formula 74) is resistant to oxime therapy and Berry and Davies (1970) made a study of the use of carbamates in protecting various species from its lethal effects.

In preliminary studies with guinea pigs it was established that the maximum symptom-free dose of physostigmine was 0.16 mg/kg in the absence of, and 0.63 mg/kg in the presence of, atropine. In fact the protective action of physostigmine varied little between these doses and so in all further studies the smaller dose was used. Results are summarised in Table 4.1 from which it is apparent that physostigmine given 10 minutes before soman, providing it was given in conjunction with

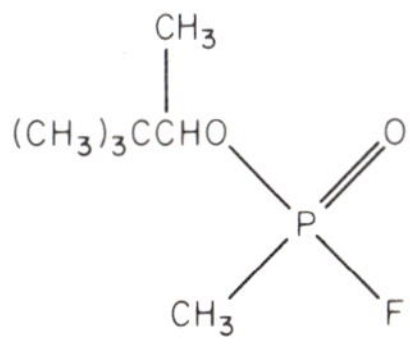

74

atropine, had an appreciable protective action. The apparent additional protection afforded by the oxime P2S (protection ratio 10.7 rather than 8.2) was not, in fact, statistically significant.

TABLE 4.1

The protective actions of various drugs and drug combinations given intramuscularly to guinea-pigs 10 minutes before soman injected subcutaneously (after Berry and Davies, 1970).

Drug or Drug Combination	Protection ratio
None	—
Atropine 17.4 mg/kg	1.5
P2S 30.0 mg/kg	1.0
Physostigmine 0.16 mg/kg	1.8
Atropine + P2S	1.8
Atropine + Physostigmine	8.2
Atropine + P2S + Physostigmine	10.7

P2S = Pyridine-2-aldoxime methyl methane sulphonate
Protection ratio = LD50 of treated animals: LD50 of untreated animals.

Physostigmine was found to be of no value if given after poisoning as might have been expected. Experiments in which the times of administration of the various drugs were varied, showed that maximum protection occurred when physostigmine was given 10 minutes pre-poisoning and atropine 1 minute post-poisoning.

Studies in other species, the results of which are given in Table 4.2, showed considerable variation, with guinea pigs, rabbits and dogs responding well but chickens, mice and rats being resistant to treatment.

TABLE 4.2

The protective actions of atropine and physostigmine given 10 minutes before soman in various species (after Berry and Davies, 1970).

Species	Dose (mg/kg) Physostigmine	Atropine	Protection ratio
Rat	0.10	17.4	2
Chicken	0.16	1.0	ca 2
Mouse	0.10	17.4	2.75
Rabbit	0.10	17.4	ca 4.5
Dog	0.10	5.0	ca 6
Guinea-pig	0.16	17.4	8.2

A number of other so-called reversible inhibitors were also studied in guinea pigs. None of them conferred significantly better protection than physostigmine although with both neostigmine and 3-(2-dimethylamino-ethyl) phenyl-N-methylcarbamate used in conjunction with atropine the protection ratio was about 7.

Berry and Davies consider that a major factor governing the effectiveness of the reversible inhibitor is the rate at which breakdown of the enzyme inhibitor complex occurs. Carbamylcholine, for example, which had very little protective action is known to produce a carbamylated enzyme which breaks down more rapidly than the mono- or di-methylcarbamates formed by the other compounds included in this study.

This kind of approach then appears to hold out some promise for prophylaxis against oxime-resistant anticholinesterase agents but there is, at present, no indication as to how man as a species would respond to such a procedure.

4.3.2 Reactivation of the enzyme

Any drugs which effectively destroy the inhibitor before it can attack the enzyme or which protect the enzyme from inhibition need to be present when poisoning occurs; that is, their use would have to be prophylactic rather than therapeutic. This presupposes some expectation that exposure to anticholinesterase agents is likely to occur. A much more likely occurrence is that poisoning will occur in an unprotected individual who will then require treatment. In such an individual an enzyme inhibitor complex will already have been formed and will need to be broken down to release an active enzyme.

The first indication that chemical reactivation of organophosphate-inhibited acetylcholinesterase could be effected is to be found in the work of Wilson (1951) who showed that choline and hydroxylamine produced a faster reactivation than water alone. In this and subsequent studies Wilson (1951, 1952, 1955) found that diethylphosphoryl-acetylcholinesterase, the enzyme inhibited by tetraethylpyrophosphate (TEPP; see formula 75), at pH7 and 31° was reactivated by 60 per cent in 4 hours by 0.25 M choline and by 50 per cent in 30 minutes by 0.1 M hydroxylamine. The ease with which reactivation could be effected depended on the nature of the inhibitor; diisopropylphosphoryl-acetylcholinesterase, the enzyme inhibited by dyflos (see formula 66) for example, was much more difficult to reactivate than the diethylphosphorylated enzyme. The activation energy for reactivation by choline was more than twice that for reactivation by hydroxylamine and consequently compounds related to hydroxylamine were studied in attempts to find more effective reactivators. Wilson and Meislich (1953) soon found that various substituted hydroxamic acids, in particular nicotine-hydroxamic acid methiodide (see formula 76), were more effective than hydroxylamine.

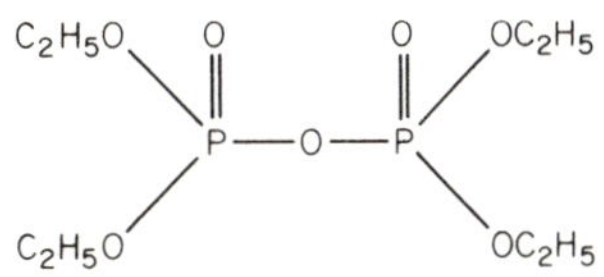

75

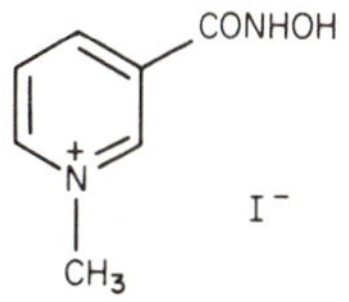

76

This compound effected a 40 per cent reactivation of diethylphosphorylated acetylcholinesterase in 15 minutes (corresponding figures for hydroxylamine: 40 per cent in 30 minutes) and a 38 per cent reactivation of the diisopropylphosphorylated enzyme in 4 hours (hydroxylamine: 17 per cent in 4 hours).

Soon came reports (for example, Davies and Green, 1955; Wilson and Ginsburg, 1955) of even more effective reactivators—the oximes. In particular these workers studied pyridine-2-aldoxime methiodide (2-PAM, pralidoxime; see formula 77) and subsequently much work has been carried out with the corresponding methyl methane sulphonate which has become known as P2S.

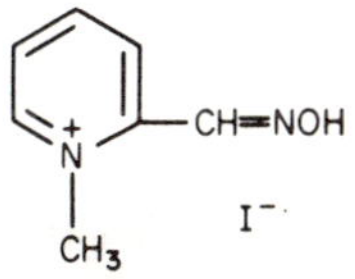

77

Also studied at about the same time were certain oxo-oximes (Childs, Davies, Green and Rutland, 1955; Holmes and Robins, 1955) of general formula RCOCR'NOH which also reactivate phosphorylated acetylcholinesterase. These include diisonitrosoacetone (DINA; see formula 78), monoisonitrosoacetone (MINA; see formula 79) and diacetylmonoxime (DAM; see formula 80).

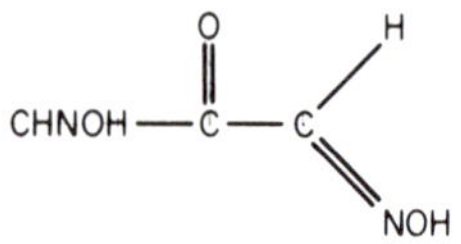

78

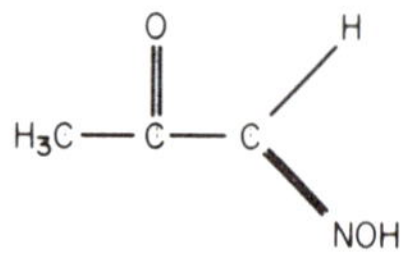

79

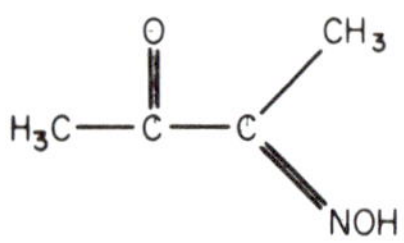

80

The bis-quaternary pyridine aldoximes (see formula 81) constitute another important, and indeed the most potent, group of oximes.

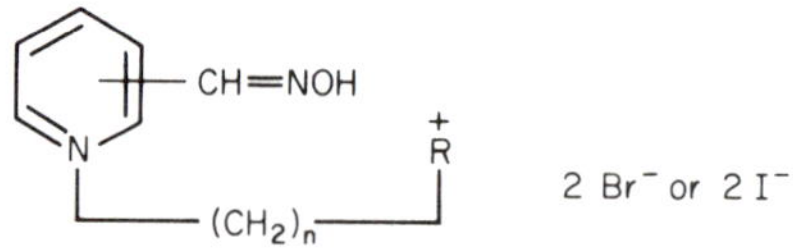

81

These were first synthesised by Poziomek, Hackley and Steinberg (1957). A particularly potent member is 1,3-bis-(pyridinium-4-aldoxime) propane dibromide (TMB-4; see formula 82) which was shown by Poziomek *et al.* and Hobbiger, O'Sullivan and Sadler (1958) to be a more rapid reactivator of sarin-inhibited cholinesterase *in vitro* than either 2-PAM or P2S.

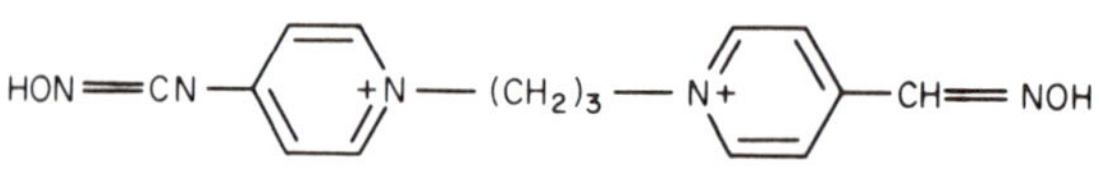

82

More recently Englehard and Erdmann (1963) reported an even more effective reactivator, bis-(4-formyl N-methylpyridinium oxime) ether bichloride (H-6, toxogonin; see formula 83).

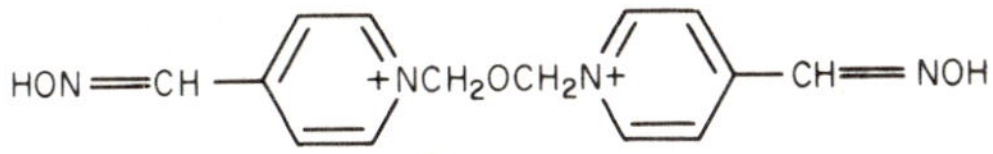

83

Thus, a whole range of substances are known which are effective reactivators of phosphorylated acetylcholinesterase *in vitro*. The above account is in no way comprehensive; for mor3 detail, reference should be made to Hobbiger (1963) where a fuller account is also given of the mechanism of reactivation by hydroxamic acids and oximes. Essentially they all act as nucleophilic reagents. A typical chemical pathway for the reactivation of dyflos-inhibited acetylcholinesterase by 2-PAM or P2S is shown in Fig. 4.2.

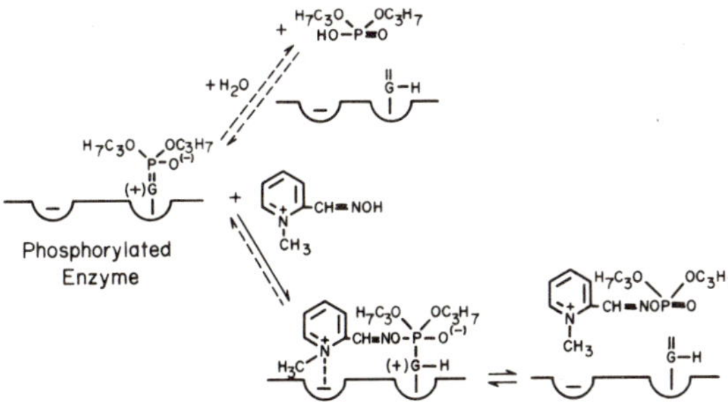

FIGURE 4.2 *Chemical pathways for the reactivation of DFP-inactivated acetylcholinesterase (AChE) by pralidoxime (below) and by spontaneous hydrolysis (above). The rate of the latter reaction is insignificant.*

Symbols used are as in figure 4.1.

Goodman and Gilman, 1970. (Reprinted with permission of The Macmillan Co., New York.)

A vast amount of work has also been carried out on the effectiveness of these reactivators *in vivo* in protecting animals from the lethal effects of organophosphorus anticholinesterase agents. Again, the field prior to 1963 was reviewed by Hobbiger. Here reference will be made only to some of the more comprehensive early work and to some later papers. It is difficult to give any really general account. There are differences between various reactivators and inhibitors and the experimental conditions which have been used are almost infinitely variable in terms of dose and route and time of administration. When the considerable difference in response between species is also considered, the picture which emerges is a very complex one.

The discussion here will be confined to oximes; hydroxamic acids are not particularly effective *in vivo* and have not been widely studied. The first point to be made is that there is no simple relationship between the *in vitro* reactivating activity of oximes and their *in vivo* potency as antidotes. Hobbiger (1963) summarised the main characteristics of this relationship as follows:

(1) Reactivating potency can be used to give an approximate indication of antidotal potency against a given organophosphorus inhibitor only when closely related oximes are used.

(2) Oximes like MINA with a high lipid solubility are more effective antidotes of the lipid-soluble inhibitor sarin than are oximes with low lipid solubility such as quaternary pyridine aldoximes.

(3) The antidotal activity of an individual oxime against organophosphorus compounds which form the same type of phosphorylated acetylcholinesterase varies greatly.

(4) The antidotal activity of oximes against organophosphorus compounds which are converted *in vivo* into phosphorylating agents is generally lower than that against compounds which are phosphorylating agents in their own right but form the same type of phosphorylated enzyme as the former.

Oximes generally are maximally effective as antidotes when they are injected some time (usually 5 to 15 minutes) before the organophosphorus anticholinesterase.

In some early work Askew (1956) studied the protective effect of five oximes against sarin (see formula 68) poisoning in the rat. The results are summarised in Table 4.3. In this study, sarin was injected subcutaneously 15 minutes after the intraperitoneal injection of the oxime. The oximes used were MINA (see formula 79), DAM (see formula 80), 2-methyl-3-oximino-4-oxopentane (see formula 84), 2-oxo-3-oximinopentane (see formula 85) and isonitrosodiethylketone (see formula 86).

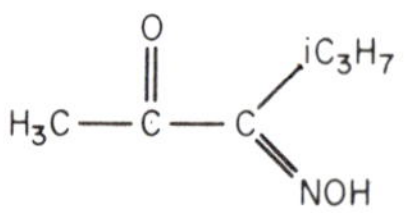

84

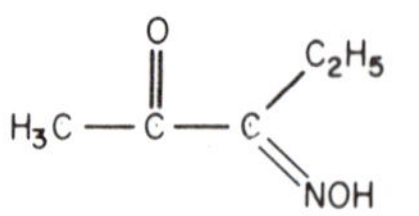

85

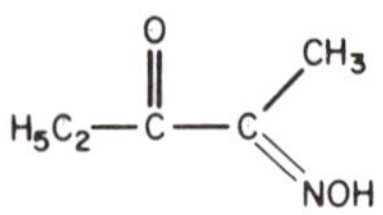

86

TABLE 4.3

Protective effect of oximes, given in equimolar doses, against sarin poisoning in the rat (after Askew, 1956).

Oxime	Dose (mg/kg)	Protection Ratio
MINA	35	4.6
DAM	41	5.6
2-Methyl-3-oximino-4-oxopentane	52	4.6
2-Oxo-3-oximino-pentane	46	5.0
Isonitrosodiethylketone	46	3.4

The results showed that these oximes elevated the LD50 of sarin by a factor of between 3 and 6.

Some further experiments were carried out with MINA and DAM. It was found, for example, that increase in dose of these oximes led to an increased degree of protection. Using the same experimental conditions as were described above, Askew found that a dose of 150 mg/kg DAM conferred a protection ratio of 26.5 while a dose of 250 mg/kg elevated the LD50 of sarin 58 times. MINA is more toxic than DAM, the maximum dose which can be given to rats without producing signs of poisoning is 35 mg/kg. This dose gave a protection ratio of between 5 and 6.

These two oximes were also studied for their effectiveness when used therapeutically, that is after poisoning with sarin. When given 30 seconds after sarin, 150 mg/kg DAM gave a protection ratio of about 14 while 35 mg/kg MINA protected from 5 to 6 LD50s.

Askew also studied the effectiveness of DAM against sarin poisoning in species other than the rat. The dose used was 150 mg/kg given intraperitoneally 15 minutes before sarin injected subcutaneously (intravenously in the monkey). The results, summarised in Table 4.4, show that, of the species studied, the rat is by far the most successfully protected by the oxime. The reason for this species difference is not clear. Askew explored the possibility that it might be related to differences in rate or degree of absorption of the oxime following intraperitoneal injection. That this was not the case seemed to be clearly demonstrated by the fact that in rabbits, blood levels of oximes were higher than those in rats in circumstances where the protection afforded was not nearly so good.

In a subsequent paper Askew (1957) reported studies with 2-PAM (see formula 77). Using a dose which produced no obvious signs of poisoning (100 mg/kg) in both rats and mice and the same kind of experimental procedure as above, she found that the protection ratios against sarin were only 1.2 to 1.3 or inferior to those obtained with DAM or MINA. The probable reason for this is that 2-PAM being a quaternary ammonium compound does not readily penetrate biological barriers. In particular it would not be expected to gain access to the

TABLE 4.4

Protective effect of DAM (150 mg/kg) against sarin poisoning in different species (after Askew, 1956).

Species	Protection Ratio
Rat	26.5
Mouse	1.7
Guinea-pig	2.0
Rabbit	1.6
Monkey	3

central nervous system. Direct confirmation of this was obtained by Hobbiger (1957a, b) who showed that a dose of 25 mg/kg 2-PAM had very little effect on the phosphorylated acetylcholinesterase in the brains of mice injected with the inhibitors tetraethylpyrophosphate (TEPP, see formula 87) or diethyl p-nitrophenylphosphate (E600, paraoxon; see formula 88) although the oxime effectively reactivated the enzyme inhibited by these substances in brain homogenates.

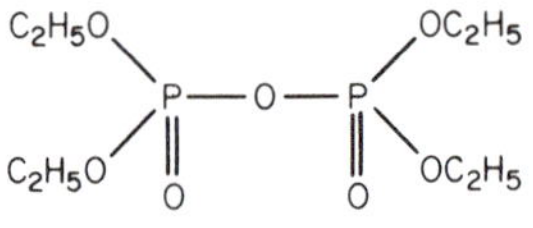

87

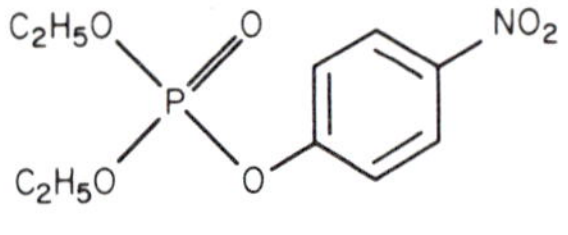

88

Another study of 2-PAM and DAM was reported by Edery and Schatzberg-Porath (1958). These authors used mice poisoned with either TEPP, dyflos, bis (dimethylamido) fluorophosphate (dimefox; see formula 89) or paraoxon (see formula 88).

In these experiments the oximes were injected intraperitoneally

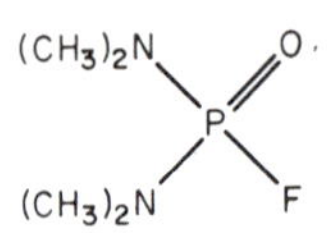

89

1 minute after the subcutaneous injection of the organophosphorus compound. Residual levels of cholinesterase were measured in the blood and brain of animals which died from or survived the poisoning. One interesting finding from this study was that 2-PAM and DAM acted synergistically in protecting mice from TEPP poisoning. Thus, while neither 50 mg/kg 2-PAM or 200 mg/kg DAM conferred any protection against 2.3 LD50s of TEPP, these two doses given together saved 18 out of 20 mice. A similar synergism was seen against dimefox but not against either paraoxon or dyflos. The second finding of significance was that blood and brain cholinesterase values in dead and surviving animals showed practically no difference, a finding lending weight to the earlier suggestion by Hobbiger (1957) that the antidotal action of 2-PAM might not be entirely due to its ability to reactivate inhibited cholinesterase.

The general conclusion to be drawn from these early studies is that a number of oximes are available, all of which are effective reactivators of inhibited acetylcholinesterase *in vitro* and all of which are capable of protecting animals *in vivo* from the lethal effects of about 1 to 5 LD50s of a variety of anticholinesterase agents. The precise degree of protection varies with the oxime, the inhibitor, the species and the exact experimental conditions used. Other considerations, notably those related to the toxicity of the oxime itself also apply in making the choice of an oxime for human use but it has been generally accepted that pyridine-2-aldoxime or pralidoxime, either as the methiodide (2-PAM, see formula 77) or the methyl methane sulphonate (P2S) is the compound of choice (Sim, 1965). More recently, however, the claims of toxogonin (Lü H-6, see formula 83) have received some support.

A comparative study of 2-PAMCl and toxogonin both from the point of view of their antidotal effectiveness and their toxicity was made recently by Fleisher, Harris, Miller, Thomas and Cliff (1970). The intramuscular LD50 of 2-PAMCl was 150 and 168 mg/kg to rats and guinea-pigs respectively. The corresponding figures for toxogonin were 225 and 79 mg/kg. Intravenously in the rat 2-PAMCl had an LD50 of 96 mg/kg and toxogonin an LD50 of 147 mg/kg. So, in the rat the toxicity of toxogonin is less than that of 2-PAMCl while in the guinea-pig the order is reversed. In rats the degree of protection against sarin (protection ratio of 1.2) afforded by both oximes was identical when they were given in intramuscular doses of 7.5 mg/kg 1 minute after subcutaneous injection of sarin. In similar experiments with guinea-pigs toxogonin conferred no protection while 2-PAMCl had a protection ratio of 1.2. In this series of experiments 2-PAMCl was also tried prophylactically when given in an oral dose of 30 mg/kg 1 hour before sarin. In rats no protection was conferred but in guinea-pigs a protection ratio of 1.3 was obtained.

During *in vitro* experiments toxogonin was rather less effective than an equimolar concentration of 2-PAMCl in reactivating guinea-pig erythrocyte acetylcholinesterase inhibited by sarin. Its approximately equal activity during *in vivo* protection experiments is presumably due, in part, to some action other than reactivation of the enzyme. This is likely to be an anti-acetylcholine action similar to that reported for the related TMB-4 by Hobbiger and Sadler (1959) and Fleisher, Moen and Ellingson (1965).

Another comparison of the effectiveness of 2-PAM, TMB-4 and toxogonin was made recently by Jovic and Boskovic (1970). This is of particular interest because these authors used as inhibitors some highly toxic phosphorylthiocholines and related compounds. In particular they studied 0-ethyl-S-(2-(N-methyl-N-phenylamino) ethyl) methyl thiophosphonate (see formula 90) and 0-(2-(N-methyl-N-phenylamino) ethyl) methylfluorophosphonate (see formula 91).

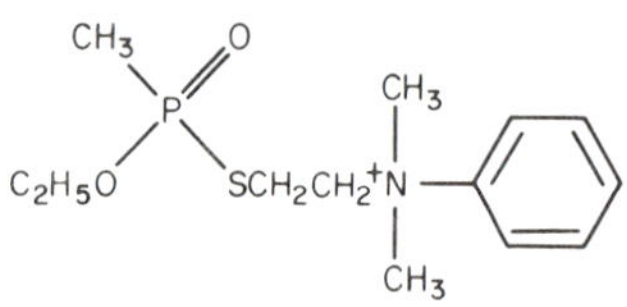

90

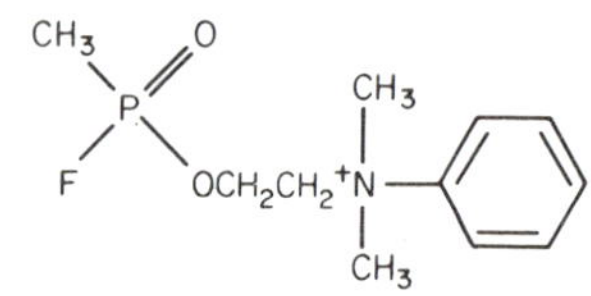

91

The subcutaneous toxicities to mice of these two compounds were 0.05 and 0.16 mg/kg respectively. The degree of protection afforded by the three oximes (0.2 LD50s) is given in Table 4.5. Against compound 90 all three oximes gave a significant degree of protection with toxogonin being the most effective. Against compound 91 only 2-PAM gave any protection; toxogonin was without effect while TMB-4 significantly increased the toxicity of the compound.

Even from the limited amount of evidence given here there is good

TABLE 4.5

Effect of 2-PAM, TMB-4 and toxogonin on the subcutaneous toxicity of compounds 90 and 91 to mice (after Jovic and Boskovic, 1970).

Oxime	Protection Ratios	
	Compound 90	Compound 91
2-PAM (48 mg/kg)	4.0	1.9
TMB-4 (17.6 mg/kg)	6.0	0.6
Toxogonin (30 mg/kg)	11.8	0.95

reason to suppose that oximes exert effects other than reactivation of the inhibited enzyme which in some situations might be helpful in protecting from the lethal actions of organophosphorus compounds. Wills (1970) mentions two possibilities—an alteration in the distribution and excretion of the phosphorylating moiety of the organophosphorus compound or a decreased incorporation of the phosphorylating moiety into the structures of functional cells of critical importance. A certain amount of experimental evidence exists to support both possibilities. For example Polak (1963) found that DAM and MINA decreased the concentration of P^{32} from labelled sarin in the blood but increased the concentration in the liver. Ågren and Ramachandran (1963) used P^{32} labelled dyflos and found that both P2S and TMB-4 reduced the rate of uptake of the isotope into the nuclear, mitochondrial, microsomal and supernatant fractions of homogenates of rat livers.

A third possibility also exists, that the oximes exert actions at the neuromuscular junction which may be, to some extent, effective in relieving the neuromuscular blockade brought about by the accumulation of acetylcholine. The fact that oximes, particularly 2-PAM which has been most studied, possess both facilitatory and depressant actions at the neuromuscular junction has been undoubtedly proved by the work of a number of authors (for example, Bethe, Erdmann, Lendle and Schmidt, 1957; Fleisher, Corrigan and Howard, 1958; Grob and Johns, 1958). The two effects are due to anticholinesterase and depolarising actions respectively.

In 1968 Goyer made a careful study of the actions of 2-PAM on neuromuscular transmission *in vitro* and *in vivo*. The oxime had an inhibitory action on the amplitude of contraction of the isolated rat diaphragm, the effect being much more marked in the indirectly than in the directly stimulated preparation indicating that the main action was at the neuromuscular junction; that is, 2-PAM had a curare-like action. No facilitating effects were observed even at low doses when there was appreciable inhibition of acetylcholinesterase. In rats the effect of the oxime was studied on sciatic nerve-tibialis anterior muscle *in situ* and the biphasic action was clearly demonstrated. Doses of up to 0.6 μmole/kg gave a small increase in contraction height but with higher doses a transient curare-like depression, roughly related to dose, was obtained. The dual action was even more apparent with the rabbit sciatic nerve-soleus and gastrocnemius muscles *in situ*. A small intra-arterial dose of 0.29 μmole produced fasciculations and potentiation of the muscle twitch. A higher dose of 58 μmoles produced a neuromuscular block. An intermediate dose of 2.9 μmoles produced little effect, presumably because the anticholinesterase action was sufficient to antagonise the neuromuscular block. Measurement of the anticholinesterase activity of 2-PAM showed that it had an I_{50} of 4 mM (compare physostigmine, 0.2 μM, d-tubocurarine 0.25 mM).

Later, Goyer (1969) made a study of the action of 2-PAM on the rectus abdominis muscle of the frog *in vitro*. Again the biphasic action was seen and confirmation that the facilitating effect was a consequence of anticholinesterase action was obtained. However, the depressing or "d-tubocurarine-like" activity was claimed to involve a non-competitive rather than a competitive inhibition mechanism.

In summary the oximes are effective *in vivo* as antidotes for poisoning by most organophosphorus anticholinesterases mainly by virtue of their ability to reactivate the inhibited enzyme. *In vitro* reactivating activity cannot necessarily be extrapolated to the *in vivo* situation due in part to the fact that the oximes have other actions but more particularly because differences exist between species, agents, routes of administration and experimental conditions.

One other fact should be mentioned here related to the phenomenon known as ageing. This describes the process whereby inhibited enzymes become resistant to reactivation by oximes or hydroxamic acids. There is still dispute as to the mechanism whereby this ageing is brought about but it seems fairly certain that at least in some cases it results from dealkylation. This was first suggested by Oosterbaan, Jansz and Cohen (1961) and has since received support from other workers. For example, Smith and Usdin (1966) represented the ageing of sarin-inhibited cholinesterase thus

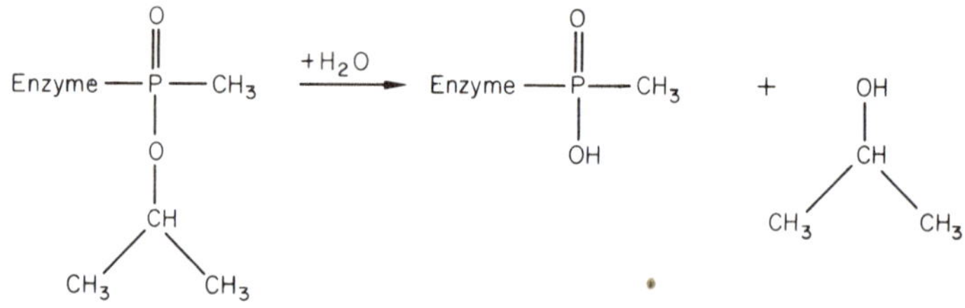

Others have suggested alternative mechanisms. For a detailed discussion reference should be made to Usdin (1970).

In the context of the treatment of poisoning with anticholinesterases this is clearly a matter of great importance particularly when, as is the case with some inhibitors, ageing occurs rapidly. Such inhibitors are virtually completely resistant to treatment with oximes. This is specifically the case with soman (see formula 74) (Loomis and Salafsky, 1963). Berry, Davies and Rutland (1966) showed that a combination of 2-PAM and atropine had little or no protective action in rats poisoned with this agent. The system was examined in more detail by Loomis and Johnson (1966) who showed that *in vitro* about 50 per cent of the soman-inhibited enzyme was not capable of reactivation by TMB-4 within 2 minutes. *In vivo* experiments with the anterior tibialis muscle of the rat showed that reactivation of the enzyme and complete recovery of function could only be obtained if the oxime was injected intravenously within 5 minutes of the intravenous injection of soman. The oxime was injected in dimethyl sulphoxide which seemed to enhance the transfer of the oxime to its site of action.

Possible methods for the treatment of poisoning with oxime-resistant agents like soman will be discussed later.

4.3.3 Antagonism of the effects of acetylcholine

During poisoning with anticholinesterase agents acetylcholine accumulates at peripheral nicotinic and muscarinic sites and also at sites within the central nervous system. No single drug can antagonise the actions of acetylcholine at all these sites. The nicotinic actions would be expected to be antagonised at the autonomic ganglia by ganglion blocking drugs and at the neuromuscular junction by curare-like agents. The muscarinic and apparently also the central actions, are antagonised by antimuscarinic or antiacetylcholine drugs. The antidotal actions of each type of drug will be considered in turn.

(i) *Ganglion-blocking drugs*

Ganglion-blocking drugs are unlikely to have any significant effect when used alone in treating poisoning with organophosphates since they act at only one of the sites where acetylcholine accumulates. In most reported studies they have been used in conjuction with atropine. Wills in 1963 reported the use of amobarbital (a barbiturate) procaine amide (a local anaesthetic) and propantheline (a quaternary amine) given to rabbits in combination with atropine sulphate 2 minutes before sarin. Only amobarbital showed any marked protective action but it is difficult to say whether this was in any way related to its ganglion blocking activity. At the dose used (20 mg/kg intravenously) there would have been considerable central nervous system depression which in itself tends to antagonise the lethal effects of anticholinesterases. Rather better protection was reported by Parkes and Sacra (1954) who found that hexamethonium used with atropine resulted in a protection ratio of about 10 compared with a figure of 2 for atropine alone.

An example of the use of hexamethonium used in conjunction with oximes is found in the study by Jovic and Boskovic (1970) who found that hexamethonium either used alone or together with TMB-4 had no protective action in mice poisoned with 0-(2-(N-methyl-N-phenylamino) ethyl) methyl-fluorophosphonate (see formula 91). When the ganglion-blocker was used with 2-PAM, however, the protection index was 2.0. Fleisher *et al.* (1970) used the secondary amine ganglion blocker mecamylamine and found that when given together with atropine 1 minute after sarin to rats at a dose of 1 mg/kg it gave a protection index of 1.9. This was increased to 2.9 if 2-PAM was also used. Mecamylamine is capable of penetrating into the central nervous system and it is possible that some of its protective action is central. Certainly, it has been shown (Bay and Adams, 1968) to reverse the actions of acetylcholine on nicotine-sensitive receptors in the medullary respiratory centres of cats which have received anticholinesterase agents.

(ii) *Curarising agents*

Non-depolarising neuromuscular blockers or curare-like agents tend to antagonise (albeit with some difficulty) the neuromuscular block caused by accumulated acetylcholine or other depolarising substances. Since this neuromuscular block is one of the main contributory factors to death in poisoning with anticholinesterase agents, it might be expected that curare-like drugs would have some value as antidotes. Wills (1963) reported that both d-tubocurarine (see formula 92) and gallamine triethiodide (see formula 93) enhanced the rate of recovery of skeletal muscle from neuromuscular blockade induced by sarin and, when used in conjunction with atropine in sarin-poisoned rabbits, these drugs gave protection ratios of 1.3 and 2.0 respectively. In this series of experiments the best protection

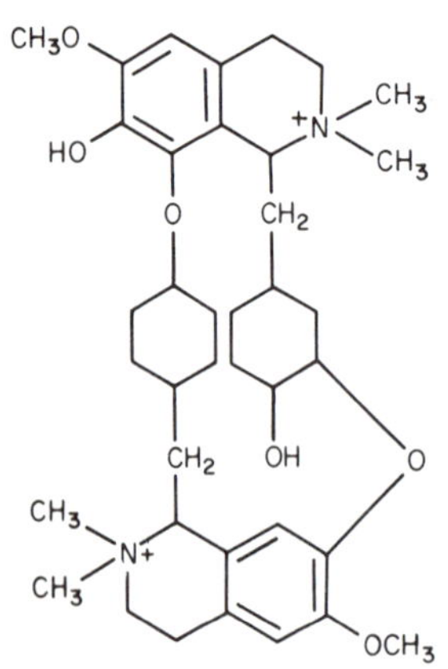

92

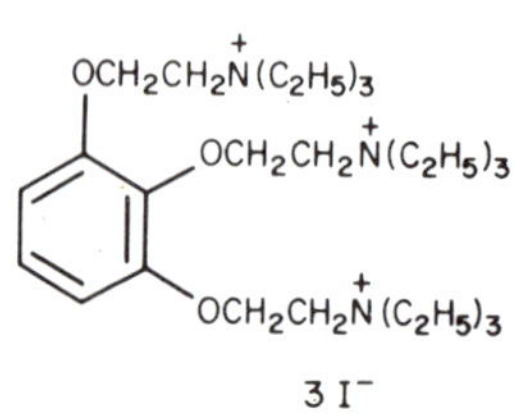

93

ratio (3.0) was obtained with ambenonium (see formula 94) an oxamide derivative with both curare-like and anticholinesterase activity.

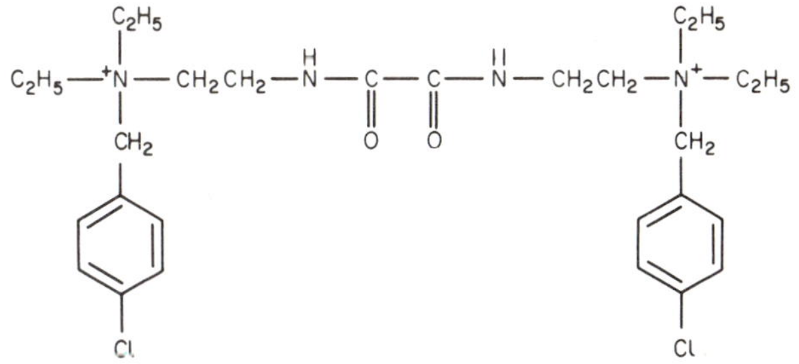

94

The above brief summary is sufficient to show that drugs which antagonise the nicotinic actions of acetylcholine, whether at the ganglion or at the neuromuscular junction, seem not to be particularly effective as antidotes to organophosphate poisoning even when used in conjunction with an antimuscarinic drug and an enzyme reactivator.

(iii) *Antiacetylcholine drugs*

The first reported use of an antiacetylcholine drug as an antidote of an anticholinesterase agent was in 1864 when Kleinwachter described the use of atropine for the treatment of poisoning with physostigmine in man. Since then there have been numerous reports of the effectiveness of atropine and other antiacetylcholine drugs in man and in experimental animals. The role of the antiacetylcholine drug is two-fold in that it tends to prevent or reverse the actions of accumulated acetylcholine at peripheral muscarinic sites but also, providing it can penetrate into the central nervous system, that is, providing it is a tertiary rather than a quaternary amine, it reverses certain central effects. These central effects include convulsions, reflex activities and the "adversive syndrome" or compulsive turning (for references see Wills, 1963) but most important is the depressant action on the respiratory centre. The ability of atropine and related drugs to reverse central respiratory depression can be demonstrated most strikingly in experimental animals. Fig. 4.3 shows a recording of diaphragm contractions taken from an anaesthetised rabbit using the "Head slip" technique (Head, 1889). Sarin given intravenously blocked spontaneous contractions but atropine rapidly restored them. Action potentials recorded from the phrenic nerve in the same preparation showed a concurrent cessation and renewal indicating that the failure was in this instance primarily central.

Peripherally, antiacetylcholine drugs reverse hypotension, bradycardia, excess glandular secretions and smooth muscle contractions and ophthalmic effects. Reversal of these effects is clearly of considerable importance in the antidotal activity of antiacetylcholine drugs but it seems likely that central actions, particularly in reversing blockade of the respiratory centre, are likely to be of paramount importance. It is not easy, however, to allot relative degrees of importance to peripheral and central actions. An early attempts was made by Karczmar and Long (1958) who measured the

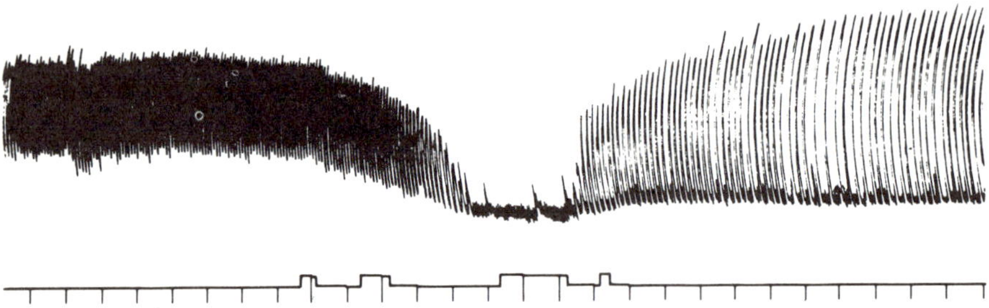

FIGURE 4.3 *Diaphragm contractions recorded from an anaesthetised rabbit.* Sarin *(25 μg/kg i.v.) at first marker caused cessation of respiration. Atropine sulphate (10 mg/kg i.v.) at third marker restored respiration. (Time Marker = 5 seconds).*

peripheral antiacetylcholine activity of thirty-five substituted tertiary amines using production of mydriasis in mice as the test method. The same compounds were then studied in the same strain of mice for their ability, when given in combination with benzoquinonium 15 minutes before a 2 LD50 dose of TEPP, to protect the animals. The tertiary amines used included alkanol substituted amines, substituted acetic acid esters, substituted tropanes, and α- and β-hydroxy acid esters. Despite this heterogeneity of compounds there was a statistically significant degree of correlation between anti-TEPP activity and peripheral antiacetylcholine potency. The implication of this finding would seem, at first sight, to be that the main, or even the sole, role of the antiacetylcholine drugs is a peripheral one. However, the results are not easy to interpret. In the first place the use of benzoquinonium complicates the situation, apparently unnecessarily, and secondly the authors also showed that the quaternized salt, atropine methyl nitrate was relatively ineffective as a protective agent when used alone and did not potentiate the protective action of the antiacetylcholine drugs. The conclusion was reached, although the experimental findings do not seem to justify it entirely, that the protection afforded by these antiacetylcholine drugs was central in nature and that the correlation between peripheral antiacetylcholine potency and protective activity suggested that the central protective mechanism involved a cholinergic system. This latter conclusion must assume a constant ratio between the peripheral and central antiacetylcholine potencies of the drugs which were used. This is highly unlikely to be the case.

There have been a number of other similar investigations with somewhat variable results. For example Dahlbom, Diamant, Edlund, Ekstrand and Holmstedt (1953) found no correlation between the peripheral antiacetylcholine potencies of a group of acyl derivatives of phenothiazines and their protective effects against tabun in the rabbit and the guinea pig. Similarly, Coleman, Little and Bannard (1962a) failed to show any correlation between the therapeutic effectiveness of a series of antiacetylcholine drugs used in conjunction with an aldoxime in sarin-poisoned mice and their peripheral antiacetylcholine activity. The situation was even more confused by findings (for example, Coleman, Little and Bannard, 1963) that, in contrast to the result of Karczmar and Long (1958), the quaternary atropine methyl nitrate does have some protective action against organophosphate poisoning when used alone or in conjunction with an oxime.

To attempt to resolve this confusion Brimblecombe, Green, Stratton and Thompson (1970) compared both the peripheral and the central antiacetylcholine potencies of a number of drugs with their effectiveness when used alone and in conjunction with the oxime P2S in protecting mice, rats and guinea pigs from the lethal effects of sarin. The results are summarised in Table 4.6.

TABLE 4.6

The protective actions against sarin of some antiacetylcholine drugs when used alone or in combination with the oxime P2S in rats, mice and guinea pigs (after Brimblecombe *et al.*, 1970).

Antiacetylcholine drug (Dose 50 μmol/kg)	Protection Ratios — Rats: Drug alone	Rats: Drug + P2S (140 μmol/kg)	Mice: Drug alone	Mice: Drug + P2S (140 μmol/kg)	Guinea pigs: Drug alone	Guinea pigs: Drug + P2S (140 μmol/kg)
Atropine (formula 9)	1.2	27.5	1.1	2.6	1.3	35.8
Hyoscine (formula 10)	1.2	9.3	1.3	2.0	1.4	21.4
Caramiphen (formula 95)	4.1	11.8	1.1	2.5	1.1	18.6
G3063 (formula 96)	9.3	79.5	1.4	6.3	1.5	58.3
PMCG (formula 97)	2.0	62.3	1.5	3.0	1.5	84.2
Ditran (formula 98)	2.0	33.6	1.4	2.4	1.4	70.2
(P2S)	1.1	—	1.2	—	1.1	—

Antiacetylcholine drugs and/or P2S were given intramuscularly 15 minutes before sarin subcutaneously.

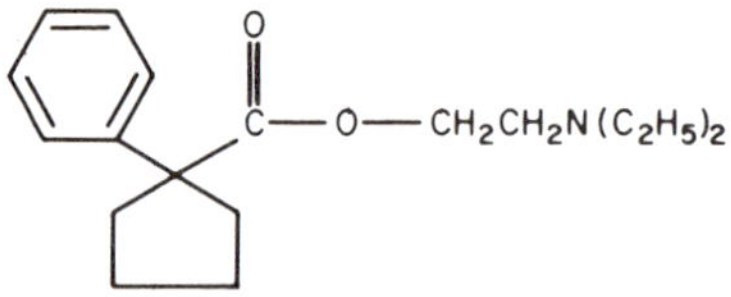

95

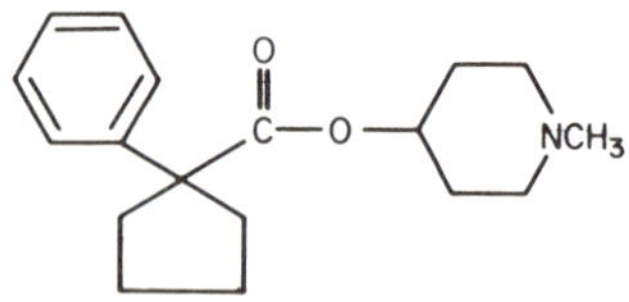

96

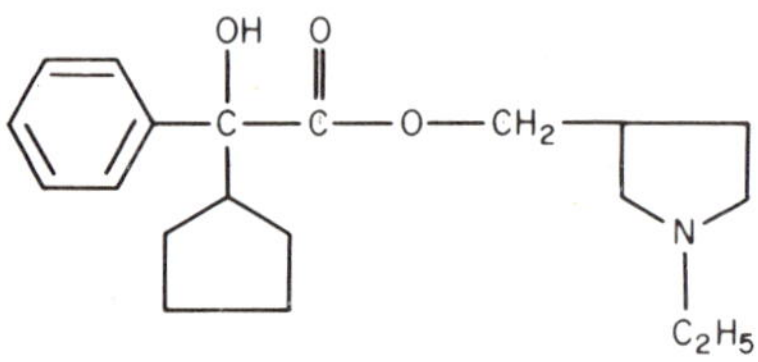

97

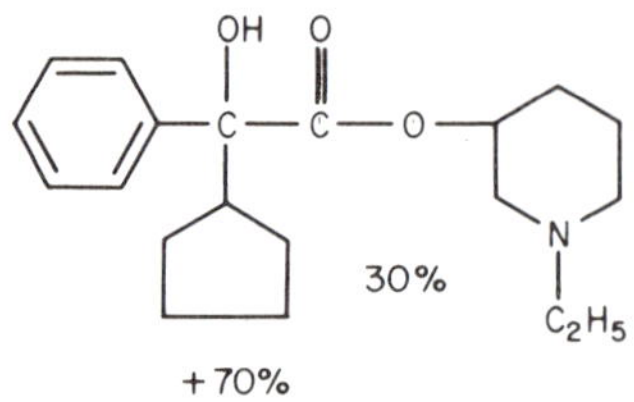

98

The results show that none of the antiacetylcholine drugs used alone gave protection ratios greater than 1.5 in mice or guinea pigs. In rats higher protection ratios were obtained with caramiphen and its analogue G3063. The addition of the oxime P2S markedly potentiated the protective actions of all the antiacetylcholine drugs used. This aspect will be discussed later in this Chapter.

One of the difficulties in a study of this kind lies in deciding on a satisfactory test for central antiacetylcholine activity. In these experiments the ability of the mice to block oxotremorine-induced tremors was used as a measure. This topic will be discussed in much greater detail in the next Chapter but at this point it suffices to say that oxotremorine (see formula 18) is a muscarinic agonist which, as well as producing peripheral parasympathomimetic effects, also produces marked muscular tremors; although there has been much discussion as to its mode of action in producing these tremors, there is now considerable evidence to support the view (Bebbington and Brimblecombe, 1965; Cox and Potkonjak, 1969a, b) that it is due to an interaction of the drug with central acetylcholine receptors analogous to peripheral muscarinic receptors. Blockade of oxotremorine-induced tremors can thus be used to give a measure of central antiacetylcholine potency. Peripheral antiacetylcholine potency can be measured in the same animals by blockade of salivation. Thus, the anticholinergic activities of all the drugs used in the protection experiments were measured using this method and by their mydriatic effect in mice. There was no statistically significant correlation between protective potency of the drugs used alone or in conjunction with P2S and any of the measures of antiacetylcholine potency.

There is the possibility that the concentrations of drugs which, in the protection experiments, were responsible for survival of the animals were different from the concentrations reached in the oxotremorine and mydriasis experiments but a more likely explanation seems to be that effects other than antagonism of acetylcholine at muscarinic receptors might play a part in protection. It is possible that the effects at the neuromuscular junction and on skeletal muscle described in the previous Chapter may be of some significance in this respect. This was studied by Brimblecombe and Everett, (1969, 1970) who were able to show that in anaesthetised cats the intra-arterial injection of the anticholinergic drug PMCG (see formula 97) protected both fast-twitch (flexor hallucis longus) and slow-twitch (soleus) muscles from the effects of sarin. The doses required were, however, high and it is not certain that equivalent concentrations would be reached in experiments where normal protective doses are used.

The general conclusion to be drawn, then, concerning the role of anti-acetylcholine drugs as antidotes to anticholinesterases is that their central actions, presumably in antagonising the actions of acetylcholine, are of greatest importance. Additionally, however, similar actions at peripheral muscarinic sites are of significance as may be other actions not normally associated with drugs of this type. Any close analysis of the relative importance of these various actions needs to be related specifically to the particular experimental conditions and cannot be applied in general terms. This was well illustrated by a study by Coleman *et al.* (1963) who investigated the effects of quaternisation and alterations of anionic function on the potency of antiacetylcholine drugs in the treatment of sarin poisoning. Both rats and mice were used and the antidotal effectiveness of tertiary amines and their corresponding quaternary salts were compared both as substitutes for, and as adjuncts to, atropine sulphate when used with P2S. The conclusion was reached that the protective capacity of the quaternary salts altered with the species used, the nature of the test, the antiacetylcholine drug, the type of alkyl group introduced and, at least in the case of caramiphen, depended on the nature of the anionic group which was present.

4.3.4 Symptomatic treatment and miscellaneous drugs

As has already been noted the central respiratory depression is a major cause of death in anticholinesterase poisoning. It is not surprising, therefore, that various central stimulants have been tested for their antidotal effectiveness. A number of workers have experimented with nikethamide (see formula 99) which, in some circumstances, appears to be beneficial. Distefano, Hurwitz, Neuman and Hodge (1951) used the drug as an adjunct to atropine in the experimental treatment of ethyl-p-nitro-phenylthiobenzene phosphonate (EPN, see formula 100) poisoning in rats.

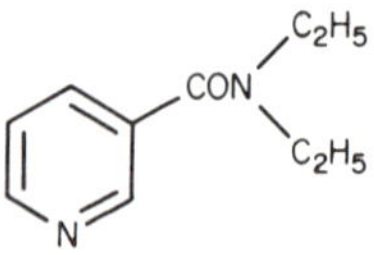

99

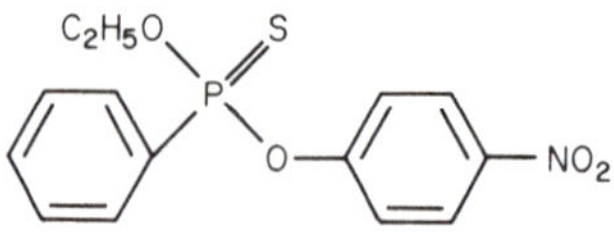

100

A sex difference emerged in the results in that the protection ratio in female rats was about six while in male rats it was less than two. In similar experiments Rosenberg and Coon (1960) also showed that nikethamide used together with atropine protected female guinea pigs and mice to some extent from EPN poisoning. Against other anticholinesterase agents it was ineffective. Rosenberg and Coon made a careful study of the role of nikethamide and discovered that any protection that it afforded was due not to the drug itself but to its metabolites especially nicotinamide which acted not as an analeptic drug but apparently as a reactivator of inhibited cholinesterase (Wilson, 1952; Rosenberg and Coon, 1960).

Wills (1963) studied other central stimulants including ephedrine (see formula 101) and pentylenetetrazol (see formula 102) used as adjuncts to atropine in anaesthetised dogs poisoned with organophosphates. He concluded that their beneficial effects were only marginal. If the administration of atropine alone was almost sufficient to restore spontaneous respiration, the addition of either ephedrine or pentylenetetrazol assured that regular spontaneous respiration persisted for several hours. When atropine only produced non-sustained, feeble respirations the stimulant drugs did not improve the condition.

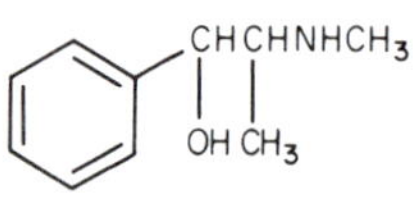

101

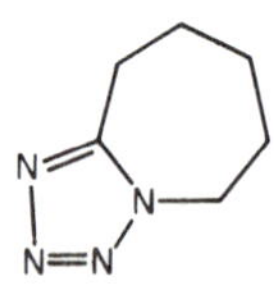

102

An important part of the syndrome resulting from anticholinesterase poisoning involves central nervous system stimulation manifested notably by convulsions and in some species of laboratory animals, by a compulsive turning, described by Freedman and Himwich (1949) as an "adversive syndrome". Attempts to counteract these stimulant effects have involved the use of a variety of central nervous system depressant drugs—sedatives, hypnotics, neuroleptics and anticonvulsants. Wills (1963) found that anticonvulsants had little value when used as adjuncts to atropine in rabbits given sarin intravenously. Sodium pentobarbital and particularly chlorpromazine did, however, potentiate the antidotal effectiveness of atropine. The protection ratio given by atropine plus chlorpromazine was about 10. It is known that chlorpromazine and indeed other phenothiazine derivatives, have weak antiacetylcholine activity but it seems unlikely that this is the reason for their effectiveness in this situation since even very large doses of atropine cannot afford protection approaching that of the combination with chlorpromazine. Wills suggests that the decrease in metabolie rate brought about by chlorpromazine plays an important role in its effectiveness. In support of this he cites the fact that the addition of sodium pentobarbital to the atropine-chlorpromazine combination increased the protection ratio still further to about 14. This may be the answer in part but it does not seem to represent a complete explanation. Experiments in the author's own laboratory have established that a roughly equivalent degree of metabolic depression brought about by other depressant drugs does not result in an equivalent degree of protection from the lethal effect of organophosphates.

It has been suggested by Wills (1970) that the anti-nicotinic actions of the phenothiazines are of relevance in this context. These actions were reported by Dahlbom, Edlund, Ekstrand and Katz (1952) who noted that a number of phenothiazine derivatives prevented nicotine-induced tremors in laboratory animals. The same derivatives were reported by Dahlbom *et al.* (1953) to have some protective effect in tabun poisoning in animals. Some kind of synergism between atropine and chlorpromazine seemed to be suggested by the results of Frada and Gucciardi (1957) who found that a combination of both was more effective than either drug alone in delaying death after poisoning with a mixture of malathion and parathion. Coleman, Little and Bannard (1962a, b) included some phenothiazine derivatives in a series of anticholinergic drugs which they studied in mice for effectiveness in the treatment of sarin poisoning. These were in no way remarkable when used in combination with the oxime P2S but triflupromazine (see formula 103) markedly potentiated the protective action of

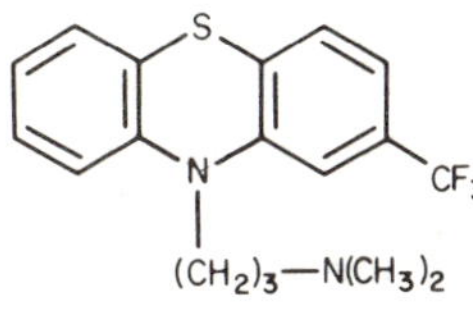

103

atropine increasing the protection ratio from about 2 to about 8. No satisfactory explanation for this potentiation can be offered at present but unpublished results from the Author's laboratory have confirmed the findings of Coleman and his colleagues and have indicated that the protective actions of triflupromazine are even more marked in the rat and the guinea pig than in the mouse.

Another example of an apparent synergism between phenothiazine derivatives and atropine will be found in the next chapter where the potentiating effects of chlorpromazine and triflupromazine upon the ability of atropine to produce a so-called dissociation between electrocortical activity and behaviour are described and discussed.

With the above findings suggesting that phenothiazine derivatives might have a useful practical role in the treatment of anticholinesterase poisoning, it is interesting to note a number of reports which, in comparison, suggest that the effects of organophosphates are potentiated rather than reduced by phenothiazines. For example Gaines (1962) found that repeated doses of both chlorpromazine and promazine increased the toxicity of parathion to male rats, reducing the LD50 from 12 to 7.3 and 5.4 mg/kg respectively. In female rats a single oral dose of 5 mg/kg of promazine increased the toxicity of a single dose of 0,0-diethyl-0-(4-nitrophenyl) phosphorothioate (parathion; see formula 104) but had no effect on the toxicity of a single dose of dimethyl-1-methyl-2-carbomethoxyvinyl phosphate (Phosdrin; see formula 105).

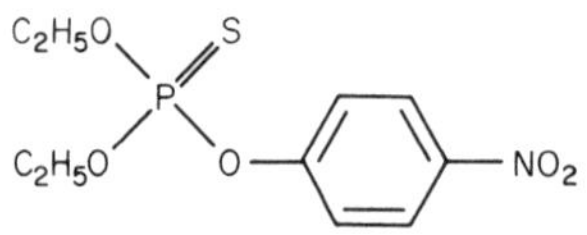

104

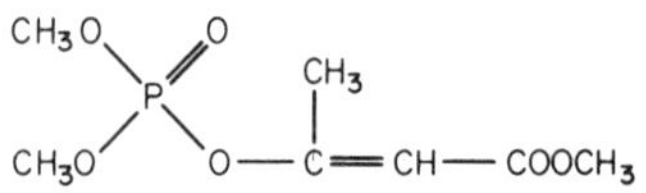

105

A human case of poisoning with Phosdrin and parathion was reported by Arterberry, Bonifaci, Nash and Quinby (1962). The patient was treated with atropine for some days and later, because he was displaying considerable restlessness, he was given a number of doses of promazine. Eventually he died. The authors concluded that the phenothiazine derivative potentiated the effects of the organophosphorus inhibitor and suggested that phenothiazine derivatives have a "paradoxical" action being beneficial in organophosphate-poisoned men and experimental animals when some active acetylcholinesterase remains but being deterimental in cases where there is complete, or almost complete inhibition of the enzyme. The detrimental effects may be due to the known weak enzyme-inhibiting properties of phenothiazines.

4.3.5 Combinations of drugs

It is clear from the discussion above that none of the drugs or types of drugs mentioned can protect from more than a few lethal doses of most anticholinesterase agents. The reason is that most antidotal drugs have fairly discrete sites of action and thus do not antagonise all the actions of the accumulated acetylcholine. The oximes, the action of which in reactivating the inhibited enzyme is more general, nevertheless show only a limited degree of protection probably because most of the more effective reactivators are in the quaternised form and thus do not penetrate into the central nervous system. Antiacetylcholine drugs while being effective in antagonising muscarinic and central actions are relatively, or perhaps completely, ineffective at nicotinic sites, particularly the neuromuscular junction. It is predictable, therefore, that a combination of antiacetylcholine drug and oxime would give appreciably better protection than either drug used alone. Many workers have demonstrated that this is the case. An example is to be found in the results given in Table 4.6 where the oxime P2S markedly potentiated the protective activity of a number of antiacetylcholine drugs in sarin-poisoned rats, mice and guinea-pigs. The degree of potentiation was much greater in rats and guinea-pigs than in mice. The reason for this species variation is not yet clear.

Clinically, then, the recommended treatment for anticholinesterase poisoning is a combination of atropine and oxime and there are now many reports of the successful use of this drug combination in human cases of poisoning, mostly with insecticides. The regimes of treatment are discussed in detail by Grob (1963). In summary, the recommendations are that the casualty should first be removed from the source of contamination and then atropine given intravenously or intramuscularly until signs of atropinisation appear (in severe cases of poisoning this may involve the administration of very large doses indeed of atropine). In severe poisoning oximes should also be given by the intravenous route with reversal of muscular weakness being used as the index of the amount of oxime to be administered. Secretions should be removed from the upper respiratory tract and a patent airway maintained, by intubation if necessary. Artificial respiration should also be applied where appropriate and convulsions, if they are not controlled by atropine, should be treated with an anticonvulsant drug like trimethadion.

The difficulty remains of treating poisoning with those agents which form a rapidly ageing enzyme-inhibitor complex and which are thus relatively or almost completely insensitive to oxime treatment. In animal experiments very large doses of oxime given immediately after poisoning have proved somewhat effective but, apart from the practical difficulties of ensuring this rapid administration, problems may well arise from toxic effects due to the large doses of oximes. At present the best treatment for agents of this type would seem to be atropine and artificial respiration. For the future the main hope probably lies in protecting the enzyme from the inhibitor, that is using the kind of procedure described in Section 4.3.1. of this chapter.

4.4 Behavioural effects of anticholinesterases

4.4.1 Experimental animals

Although this section of the monograph is concerned specifically with responses of the whole organism to anticholinesterase agents it is worth mentioning by way of introduction that these agents have a specific effect on the EEG in producing a pattern of electrical activity characterised by low amplitude, high frequency waves (Bradley and Elkes, 1953, 1957). This effect may be associated with cholinergic activity: it can be reversed by atropine or other antiacetylcholine drugs. The relationship between this effect on the EEG and the behaviour of the organism is, however by no means clear since this desynchronisation of the EEG is normally associated with alert behaviour whereas in animals given anticholinesterase agents the change in the EEG pattern is not associated with a concomitant alerting, that is, there is a so-called dissociation between electrocortical activity and behaviour. A fuller description of the effects of anticholinesterases on the electrical activity of the brain and of various other central actions of these agents is given by Machne and Unna (1963).

Behavioural effects which have been described can be divided conveniently into effects on spontaneous behaviour and effects on behaviour involving some kind of apparatus for their detection.

One of the most commonly described spontaneous effects is a compulsive circling or vestibular or adversive syndrome first reported by Schiff, Esmond and Himwich (1950), following the intracarotid injection of dyflos to rabbits. Subsequently other workers reported a similar syndrome in other species and with other agents. Destruction of the vestibular nuclei in the brain stem prevented the occurrence of the syndrome. It has still not been established with certainty that the effects result from the anticholinesterase activities of the agents used; however, Harwood (1954) showed that the cholinesterase depletion was greater on the side of injection than in the other hemisphere. The effect reported by Schiff, Esmond and Himwich (1950) in rabbits was reversed by dimenhydrinate while a similar effect produced by the injection of dyflos into the caudate nucleus (White, 1956), and associated with depletion of cholinesterase in this nucleus, could be prevented by atropine.

Many studies of the general behaviour of animals have suggested a depressant type of action by the anticholinesterases but the cat appears to be somewhat atypical in this respect. For example, Feldberg and Sherwood (1954) reported that after intraventricular injections of physostigmine (20–100 μg) or dyflos (100 μg) cats displayed itching, certain motor phenomena and a "catatonic" stupor. These effects, they claimed, resulted from inhibition of cholinesterase and the accumulation of acetylcholine in the region of the periventricular grey matter. These authors suggested that the effects were reminiscent of phenomena seen in schizophrenic patients but this suggestion does not seem to have received much support subsequently.

There has been a good deal of work concerned with the effects of anticholinesterase agents on various types of conditioned behaviour in experimental animals. The studies previous to 1967 have been summarised by Bignami and Gatti (1967) and only a few representative investigations will be discussed here. In view of the great variety of apparatus, drug, species, doses and dose regimes which have been used it is, however, difficult to draw any general conclusions. Bignami and Gatti list a large number of references, the results from which indicate that both carbamates and organophosphorus compounds have markedly depressant effects upon previously acquired conditioned behaviour. The more interesting studies are those in which attempts have been made to correlate behavioural changes with the degree of brain cholinesterase inhibition. One such study was by Russell, Watson and Frankenhaeuser (1961) who used 0,0-diethyl-S-ethylmercaptoethanol thiophosphate (demeton-S, Systox; see formula 106) as the anticholinesterase agent. This was given in the diet at five different concentrations to rats which were trained in a conditioned avoidance response situation. Once an arbitrarily determined criterion of conditioning had been reached extinction trials were carried out in which animals were subjected to the conditioned stimulus but not the shock or unconditioned stimulus.

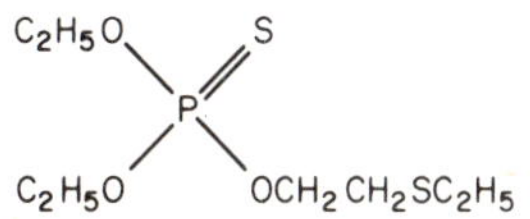

106

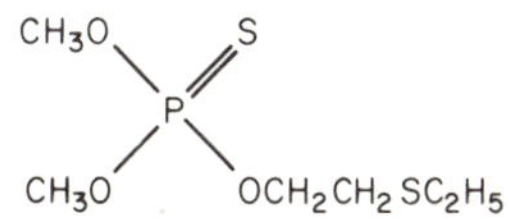

107

It was found that the five concentrations of demeton-S in the diet gave a dose-related reduction in brain cholinesterase levels ranging from a 36.6 per cent reduction in enzyme activity with 3 p.p.m. in the diet to a 76.5 per cent reduction with 40 p.p.m. No effect was seen upon speed of conditioning or acquisition of the conditioned response expressed in terms of the number of trials required to reach the criterion of conditioning.

An effect was, however, seen upon the extinction of the response in terms of the percentage of animals which extinguished the response within a limit of fifty trials. It seemed that there was a "critical level" between 5 and 10 p.p.m. of demeton-S in the diet, corresponding to 60 to 65 per cent reduction in brain cholinesterase activity below which the rate of extinction was reduced.

The implications of these experiments are interesting. In the first place it is clear that not all aspects of behaviour are affected concurrently. The speed of conditioning was unaffected in animals with reductions in brain cholinesterase of about 80 per cent while extinction of the response was clearly modified in animals with 60 per cent reduction in enzyme activity. Secondly, the concept of a "critical level" is an important one. It appears from these results that a considerable excess of cholinesterase is present in the central nervous system since more than half of it must be inhibited before behavioural effects can be detected. Eventually, however, a point is reached where the amount of enzyme present is insufficient to enable normal behaviour to be maintained. In this context Banks (1957) reported that problem-solving ability in the rat began to be affected when cholinesterase levels were depleted to a similar extent, that is by 60 to 80 per cent. Overt signs of poisoning in rats receiving demeton-S do not appear until brain cholinesterase levels are reduced by 92 to 93 per cent (Barnes and Denz, 1954).

Russell, Watson and Frankenhaeuser (1961) linked their finding to an earlier suggestion by Krech, Rosenzweig and Bennett (1956) that levels of cholinesterase in the brain reflect the readiness for nerve impulse transmission in that organ and that the ease with which nerve impulse transmission can occur is related to the capacity for adaptive behaviour. Russell and his colleagues suggested that the speed of extinction might be the pacemaker step in a series of adaptive behavioural changes. The fact that the rate of extinction is reduced in animals with considerably depleted levels of brain cholinesterase seems to lend support to the views of Krech, Rosenzweig and Bennett.

It should, however, be noted that other workers have reported increases in the rate of extinction following anticholinesterase agents and in one study by Glow and Rose (1965), a biphasic effect was seen. In rats given a single dose of dyflos (1 mg/kg intramuscularly) there was an acceleration of extinction during the first session after treatment but in later sessions extinction was retarded. However, in animals treated with a reactivator of inhibited peripheral cholinesterase the initial acceleration of extinction was slowed down while the later retardation effect was completely eliminated. The clear implication of these findings is that the retardation effect is due to a peripheral, rather than a central, action.

The methyl derivative of demeton-S (methyl demeton; see formula 107) was studied in cats by Stacek (1959, 1960). A single oral dose of 10 mg/kg reduced conditioned reflexes and on the day after administration several cats no longer showed motor responses to conditioned stimuli. At this time erythrocyte and serum cholinesterase activities were reduced by 56 to 59 and 66 to 77 per cent respectively. The extent by which brain cholinesterase levels were depleted was not measured. A lower dose of 5 mg/kg while reducing erthrocyte and serum cholinesterase levels by 15 to 20 per cent had no influence on conditioned reflexes and a dose of 2.5 mg/kg had no effect on either enzyme levels or conditioned reflexes.

Stacek also studied the effects of methyl-demeton administered chronically to cats by inhalation. When erythrocyte cholinesterase levels were depleted by 37 to 52 per cent a decrease in strength of conditioned reflexes was noted. From all his experiments Stacek considered that the "critical level" for erythrocyte enzyme inhibition was 40 per cent. In animals where the inhibition was in excess of this there were reductions in the strength of conditioned reflexes.

Kagan (1962) also studied the effects of demeton-S on cats and his results also indicated that reductions in erythrocyte cholinesterase activities of 40 to 50 per cent were accompanied by reductions in the strength of conditioned reflexes. The other interesting finding was that if demeton-S was given daily in an oral dose of 0.25 mg/kg blood cholinesterase levels remained depleted but the conditioned reflexes gradually returned to normal presumably due to some kind of compensatory process within the central nervous system. The recovery of the reflexes was accelerated by the administration of the antiacetylcholine drug caramiphen (see formula 96) presumably confirming that the effects were due to actions on cholinergic systems. Buryj (quoted by Medved, Spynu and Kagan, 1964) clearly shows that these kinds of effects on conditioned reflexes are central in origin when he used octamethyl pyrophosphortetramide (OMPA, schradan; see formula 108).

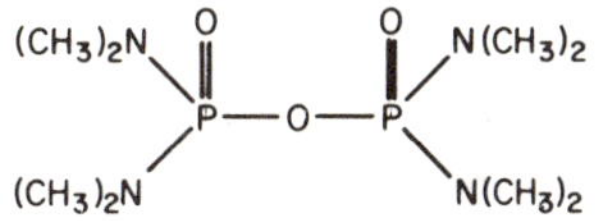

108

This compound, which has been used as an insecticide, has a mainly peripheral action and thus has little inhibiting action on brain cholinesterase. Buryj found that changes in the conditioned reflexes in animals receiving daily doses of OMPA occurred only shortly prior to death when gross overt signs of poisoning were present. Similarly Funderburk and Case (1947) used the quaternary ammonium salt neostigmine (see formula 63) in cats and found that conditioned behaviour was only disrupted when severe signs of poisoning appeared and that atropine and the quaternised methylatropine were equally effective in reversing the effects which were clearly peripheral in origin.

There have been a number of studies on the interaction of anticholinesterases and antiacetylcholine agents. For example, Michel'son, Rozkova and Cavateev (1954) showed that the inhibitory effects of atropine and hyoscine on conditioned reflexes were blocked by physostigmine and reference has already been made to the work of Kagan (1962) showing that the effects of demeton-S were reversed by caramiphen.

Although most of the studies so far referred to indicate a depressant action on conditioned responses it has been pointed out by Bignami and Gatti (1967) that there are a number of examples to indicate that anticholinesterase agents can have dual actions with low doses producing reduction of latencies and increases in the frequency and strength of conditioned responses with opposite effects being seen following higher doses.

Another set of investigations has been concerned with the effect of anticholinesterase agents on the rate of learning or conditioning rather than on the already established conditioned response. Richardson and Glow (1967a) chronically reduced brain cholinesterase activity in rats by repeated injections of dyflos and found that in these animals there was impaired visual pattern discrimination learning when a food reward was the only incentive but not when errors were also punished by shock. This result was interpreted as meaning that low cholinesterase levels affected the performance of the animals but not their ability to learn. In a subsequent paper Richardson and Glow (1967b) found a similar effect in the performance of previously learned discrimination.

Although no effect on rate of learning was detected in these experiments of Richardson and Glow there are a number of other examples to be found in the literature of anticholinesterase agents having both facilitatory and inhibitory effects on learning (see Bignami and Gatti, 1967). A study by Stratton and Petrinovich (1963) illustrates the biphasic action of physostigmine. The drug was given to Tryon maze-bright and maze-dull rats in intraperitoneal doses of 0.25, 0.5, 0.625, 0.75 or 1 mg/kg and then the animals' ability to learn a Lashley III maze was investigated. Both maze-dull and maze-bright strains acquired the ability to run the maze more quickly after low doses of the drug but higher doses resulted in a disruption of performance. The optimal dosage for facilitation of learning in the maze-dull rats (0.75 mg/kg) was higher than that for the maze-bright rats. In fact in the maze-bright rats, 0.75 mg/kg retarded learning. Stratton and Petrinovich took their results to lend support to the view that learning efficiency is related to the acetylcholine: cholinesterase ratio in the brain.

Another example of a biphasic action of anticholinesterases is to be found in the unpublished results of Russell and his colleagues (quoted by Bignami and Gatti, 1967) indicating that serial problem solving by rats is facilitated by doses of anticholinesterases producing a 40 per cent inhibition of brain enzyme. Higher doses, producing a 60 per cent inhibition or more, resulted in a decrement in performance.

Most of the investigations reported in this section so far deal with effects of anticholinesterase agents on specific aspects of behaviour. A good example of a study dealing with a variety of psychopharmacological effects of a single agent is given by the work of Goldberg, Johnson, Knaack and Smyth (1963). They studied N-methyl 3-isopropyl phenyl carbamate (Compound 10854; see formula 109).

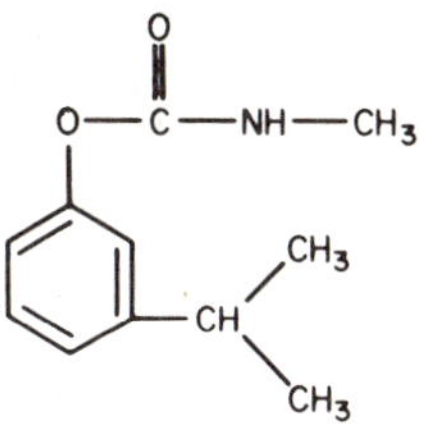

109

There was a marked sex and species variation in susceptibility to this compound. The intraperitoneal LD50 to male rats was 15.2 mg/kg and to female rats 48.0 mg/kg. The LD50 to female mice was 8.1 mg/kg.

Effects on conditioned avoidance response of female rats were studied using a pole-climb technique. The compound produced a separation of avoidance and escape behaviour, i.e. conditioned and unconditioned responses. The ED50 for block of avoidance behaviour was 1.93 mg/kg and for block of escape behaviour 5.74 mg/kg. Atropine sulphate inhibited the effects on both avoidance and escape behaviour but the peripherally acting methyl atropine only blocked effects on escape behaviour.

Discrete avoidance studies were also carried out in which rats were trained to avoid a shock by pressing a lever on receipt of a light stimulus. The effects of Compound 10854 given by inhalation were investigated. It was generated as a dust at a concentration of about 45 mg/m^3 and a mean particle size of 1.2 μ and animals inhaled the dust for either 10, 20, 30, 60 or 120 minutes. Decrement in performance was noted at all these times and showed a progressive increase with increase in time of inhalation. Brain cholinesterase levels were measured and it was found that threshold effects on this behaviour were associated with a 50 per cent inhibition in total brain enzyme activity. Similar findings came from experiments in which Compound 10854 was given by intraperitoneal injection.

Forced and spontaneous motor activity were measured using rotating rods and activity cages respectively. In each case depressant effects were noted, with 1.9 mg/kg of the compound preventing 50 per cent of the mice from remaining on the rotating rod and 2.12 mg/kg reducing spontaneous activity by 50 per cent. The compound also potentiated barbiturate-induced anaesthesia, an effect antagonised by atropine sulphate as were the inhibitory effects on avoidance behaviour.

The general picture to emerge from this study and indeed from the majority of studies reported here is one of depressant effects on behaviour brought about by anticholinesterase agents although as had already been made clear this is not always the case. It seems fairly certain that most of the described effects are due to inhibition of brain cholinesterase and therefore presumably to accumulation of acetylcholine at central sites. That many of these sites are "muscarinic" in nature seems to be implied by the finding that atropine antagonises many of the reported effects. If all these conclusions are correct it should be possible to gain some understanding of the role of central cholinergic mechanisms in the behaviour of the organism.

The aspect of behaviour which has been most studied in this context is learning and, if it is possible to draw a general conclusion from the many investigations which have been carried out, it is that high brain cholinesterase enhances learning (Rick, Morris and Kerkut, 1968). Practically all the work has been carried out on rats and it may not be entirely justifiable to extrapolate generally from this species but, in the absence of any indications to the contrary, it seems reasonable to assume that the conclusion applies equally to other species. It has been stated by Koelle (quoted by Russell, 1969) that the localisation of a high level of acetylcholinesterase in a group of neurones seems to be a reasonable indicator of cholinergic function, and it is tempting to link levels of the enzymes with levels of cholinergic activity. However, this may not be the case as Rosenzweig, Krech and Bennett (1960), although they originally assumed that cholinesterase activity gave a good index of acetylcholine activity, later rejected this when it became apparent from their breeding experiments that the two activities were under separate genetic control.

Despite these reservations it seems clear from a number of studies summarised above that once a certain degree of brain cholinesterase depletion has been achieved then behavioural changes, almost certainly associated with acetylcholine accumulation, can be detected. Russell (1969) considers that some of these changes result from an interference with the encoding of relevant information and also from a failure to release previously extinguished behaviour which is of relevance in a new situation.

A very interesting series of experiments with anticholinesterase agents which may help to elucidate the physiological basis of memory has been carried out by Deutsch and his colleagues in California. The experiments differ essentially from most of those described above and discussed by Russell (1969) in that the agents were injected after, rather than before, training and so their effects on recall, rather than on acquisition, are being studied. The work is summarised and discussed by Deutsch (1969). The early results showed that either dyflos injected intracerebrally or physostigmine intraperitoneally to rats trained to escape a shock by running to the lit arm of a Y maze had a similar effect in that they produced a small degree of amnesia when injected half an hour after training. At three days after training the drugs had no effect but at seven and fourteen days later there was again an amnesic effect. The suggestion was made that the drugs were producing their blocking effects at high levels of synaptic conductance when the increased levels of acetylcholine they produce would cause a depolarization and hence a block. At low levels of synaptic conductance, on the other hand, additional acetylcholine would facilitate transmission. If this suggestion is correct then it can be assumed that, with the exception of the period immediately after learning, there is a progressive increase in synaptic conductance and hence an increase in susceptibility to blockade by anticholinesterase agents. To test this idea hyoscine, an antiacetylcholine drug and one which reduces the effective concentrations of acetylcholine in the brain, was used. As might have been expected from the theory, the drug blocked memory at the time when the anticholinesterases had no effect and when conductance was presumably low. On the other hand it had no effect at the time when anticholinesterases were effective in producing amnesia.

Another type of experiment was carried out in which the effect of dyflos on an almost forgotten response was studied. The drug had the effect of facilitating the response and Deutsch considers that forgetting is due to a lowering of synpatic conductance which is reversed by the anticholinesterase agent.

Although the anticholinesterase agents were without effect at 3 days after training and yet produced an amnesia at 7 days later there were no apparent differences in the performances of control animals at these times. Deutsch attributes this to a "ceiling" effect, that is, the animals were overtrained and showed a well-nigh perfect performance throughout this period. When they were undertrained it was noted that decrement in performance appeared with time and an interaction between degree of training and effect of the anticholinesterase drug has been noted, with weakly learned habits being facilitated and well learned habits blocked. Again the suggestion is that the results can be explained in terms of synaptic conductance which might be expected to be at a low level in undertrained animals and thus susceptible to potentiation by the anticholinesterase. The conductance would be at a high level in overtrained animals and thus susceptible to blockade.

Deutsch suggests that the implication of his findings is that extinction is a separate habit which can be left intact and capable of being remembered on retest while the original habit is unavailable. If this is the case, a rat preinjected with an anticholinesterase drug or extinguished soon after learning should find it relatively easy to relearn the original habit. Rats were thus trained to approach a light in a Y maze for a sugar-water reward 7 days before an injection of physostigmine. They were then placed in the maze and not given a reward for approaching the light until they reached a criterion of non-responding. The rats which were extinguished 3 days before the injection of physostigmine took much longer to relearn or retest than did controls and rats extinguished 6 days before receiving physostigmine. These rats did, however, learn a reversal of the original habit more quickly than the controls. The suggestion, then, is that a separate habit is learned during extinction which opposes performance of the habit which was initially rewarded. This opposing habit could take the form of an aversion to the initially rewarding stimulus.

These studies by Deutsch and his colleagues have been considered at some length here firstly because they represent elegant and interesting pieces of work and secondly to illustrate the way in which theories of central nervous system functions, even involving difficult and complex concepts like memory and learning, can be built up from experiments with drugs. The point should however be made, and will be made at greater length in the final Chapter of this monograph, that it is possible to take an over-simple view of drug actions. Karczmar (1969), in discussing this matter, pointed out that anticholinesterase drugs can have direct actions as well as indirect actions depending on their ability to inhibit the enzyme. Similarly, Deutsch's use of hyoscine as a drug which depletes levels of acetylcholine could have alternative interpretations since this drug also is a competitive antagonist of acetylcholine at muscarinic sites and has other pharmacological actions.

No specific criticism of the work of Deutsch is implied in these remarks. In fact his work seems a model of its kind and makes a significant contribution to the understanding of physiological processes underlying memory. Rather, the remarks are meant as general words of caution concerning the interpretation of experiments involving centrally acting drugs, the mode of action of which is not fully understood.

All the effects of drugs so far described seem to be related to inhibition of brain cholinesterase. In some studies this was measured, in other studies only blood cholinesterase levels were monitored and in others it is only possible to imply from the doses used that inhibition was likely to have occurred. Some recent evidence has suggested, however, that behavioural effects can, in some tests with some agents, be detected at times when brain cholinesterase levels are normal. The work, reported by Brimblecombe (1970) and Brimblecombe, Coult, Deane and Parkes (1971) concerns some halo-substituted vinyl phosphorus esters. These compounds, listed in Table 4.7, were shown to be inhibitors *in vitro* of both acetylcholinesterase and cholinesterase and there was a statistically significant correlation between log LD50s for these compounds and log kIn (kIn = rate of enzyme inhibition) for acetylcholinesterase indicating that their lethal effects were due to inhibition of this enzyme. The compounds were also studied for their effects on behaviour using an open-field technique as described by Brimblecombe (1963) but there was no statistically significant correlation between minimal effective doses in the open-field and rates of enzyme inhibition. The results are given in detail in Table 4.7.

In view of this finding some of the compounds were given to rats by subcutaneous injection, the route used in the open-field experiments, and blood and brain cholinesterase levels determined. The results, summarised in Table 4.8 showed that in no case was brain cholinesterase significantly depleted by the minimum dose of the compound which had a significant effect on open-field behaviour.

There are a number of possible explanations for these findings. The most obvious one is that the behavioural effects of the compounds resulted from actions other than inhibition of cholinesterase although it is always possible that the relatively crude procedure of homogenising the whole brain and determining the total enzyme activity masked significant degrees of inhibition in specific brain areas. Another possibility is that the behavioural effects are due to a peripheral, rather than a central, action. This does not seem very likely, however, because at the minimal effective open-field doses there was variable but relatively slight inhibition of blood cholinesterase and other experiments in the author's laboratory with a number of anticholinesterase agents have shown that no behavioural changes in the open-field can be detected with doses of agents giving this degree of inhibition of blood cholinesterase.

Another aspect of the behavioural actions of anticholinesterase agents

TABLE 4.7

Some halo-substituted vinyl phosphorus esters and related compounds with their rates of inhibition of acetylcholinesterase (AChE) and cholinesterase (ChE). Also shown are LD50 values and approximate minimal effective doses (M.E.D.s) in the open-field test (after Brimblecombe *et al.* 1971).

General formula

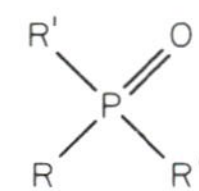

Compound No	R	R′	R″	AChE k_{In} ($l. mol^{-1} min^{-1}$) Bovine erythrocyte	Rat brain	ChE k_{In} ($l. mol^{-1} min^{-1}$)	LD50 (mg/kg s.c. rats)	M.E.D. Open-field test (mg/kg s.c. rats)
						3.1×10^5	1.25	0.02
1	C_2H_5	$Cl.CH_2CH_2O$	$Cl_2C = CHO$	3.7×10^3	4.2×10^3	1.5×10^5	2.69	0.1
2	CH_3	FCH_2CH_2O	$Cl_2C = CHO$	4.6×10^3	3.2×10^3	2.3×10^5	0.79	0.1
3	CH_3	$ClCH_2CH_2O$	$Cl_2C = CHO$	7.0×10^3	4.7×10^3	4.0×10^4	1.01	0.1
4	CH_3	C_2H_5O	$Cl_2C = CHO$	1.2×10^3	3.0×10^3	2.8×10^2	50.0	0.01
5	CH_3	C_2H_5O	$ClCH_2CH_2O$	6.9×10^1	7.5×10^1	3.5×10^5	15.5	0.2
6	C_2H_5O	C_2H_5O	$Q_2C = CHO$	1.3×10^2	2.0×10^2	5.6×10^4	13.0	5.0
7	C_2H_5O	C_2H_5O	$ClCH_2CH_2O$	1.0×10^2	8.0×10^1	3.4×10^4	35.0	0.2
8	CH_3O	CH_3O	$Cl_2C = CHO$	1.3×10^2	3.0×10^2			

TABLE 4.8

Inhibition of blood and brain cholinesterase in vivo by some of the esters listed in Table 4.7 (after Brimblecombe *et al.*, 1971)

Compound No	Dose mg/kg (s.c. injection) (* = M.E.D. in open-field test)	% inhibition of cholinesterase Brain	Plasma	Erythrocytes	
3	0.05	1	0	12	
	0.1*	0	17	31	
	0.2	5	42	62	
	0.4	31	50	73	
4	0.05	0	25	24	
	0.1*	0	24	31	
	0.2	9	55	66	
	0.4	39	78	88	
1	0.003	1	0	0	
	0.02*	2	8	0	
	0.12	5	15	23	
	0.72	58	79	86	3/8 died
6	0.125	0	7	2	
	0.5 (0.2*)	20	31	24	
	2.0	72	67	71	
	8.0	88	88	92	8/8 died

Blood and brain samples taken 90 minutes after injection or at death when this occurred earlier
All values represent the mean of 8 estimations

which warrants mention is the fact that tolerance develops to behavioural changes despite continued depression of brain cholinesterase levels. In a typical series of experiments Russell, Warburton and Segal (1969) gave rats an initial dose of 1 mg/kg dyflos followed at 3-day intervals by booster doses of 0.5 mg/kg. This maintained levels of brain cholinesterase at levels of about 27 per cent normal. Various behavioural characteristics of these animals were studied. In all cases there were significant decrements in behaviour, initially followed by recovery to control levels. The time for this tolerance to develop differed according to the nature of the behaviour being measured. Thus tolerance times were relatively short (about 10 days) for behaviours which were primarily of a peripheral motor nature, for example a conditioned avoidance response, whereas tolerance took 18 to 22 days to develop in behaviours involving so-called "central cholinergically-coded neural systems", for example water intake following water deprivation and response inhibition in an alternation situation. It was concluded that return to normal behaviour is not attributable to a recovery of brain cholinesterase activity but that two possible mechanisms are involved. One possibility is that a process of end-product inhibition occurs, that is, that the decreased cholinesterase activity results in increased levels of free acetylcholine which in turn reduce the rate of synthesis of acetylcholine. Some support for this view is provided by the work of Stavinoha *et al.* (1962) who found that daily injections of the anti-cholinesterase agent Di-Systox resulted in increases in brain acetylcholine content to a peak level after three daily injections but thereafter the levels decreased progressively until by about the tenth day they had reached the control level where they remained steady. A second possibility is that of continuous interaction of acetylcholine with its receptors with the consequent development of tolerance. This is a mechanism analogous to that proposed by Axelrod (1956) for the development of tolerance to narcotic analgesic drugs.

Subsequently Russell *et al.* (1971) compared the effects of the anti-acetylcholine drugs, atropine and methylatropine, on control rats and rats which had become behaviourally tolerant to dyflos. They found that methylatropine had little effect but that atropine caused similar decrements in performance in an operant conditioning situation in both groups until a critical point was reached. With further increase in dose of atropine the tolerant rats showed complete failure to perform whereas the control animals continued to perform at about 40 per cent normal. A possible explanation was that the receptors had become less sensitive to acetylcholine and that in a competitive situation this would favour atropine which at high doses might be expected to produce a complete functional blockade of the processes subserving the particular aspect of behaviour being studied. A similar conclusion might be drawn from the experiments of Overstreet, Hadick and Russell (1972). They studied the effects of amphetamine and pilocarpine on the eating behaviour of normal rats and rats with chronically low brain cholinesterase levels induced by regular administration of dyflos. Amphetamine reduced the food intake of both groups to a similar degree whereas the cholinomimetic pilocarpine reduced the food intake of the cholinesterase-depleted rats to a considerably smaller extent than that of normal animals. This finding could be explained on the basis of a change in sensitivity of acetylcholine receptors.

The question as to whether the short term (about 10 day) tolerance could be explained in terms of peripheral mechanisms and the longer term (about 22 days) tolerance in terms of central mechanisms was explored by Carson, Jenden and Russell (1973). It was found that isolated ilea from rats chronically treated with dyflos regained their normal sensitivity to acetylcholine after 22 days of dyflos treatment. Morphine (50 μg/ml) reduced the sensitivity of ileal strips from control rats but increased that of strips from tolerant rats. This suggested that the mechanism of tolerance might be an activation of an acetylcholine re-uptake system. These results do not, however, give any adequate explanation for the 10 day and 22 day tolerances mentioned above.

So although it is clear that tolerance can develop to the behavioural effects of chronically administered anticholinesterase agents there is, as yet, no completely satisfactory explanation for the mechanism whereby this tolerance develops. The fact that it does occur is of relevance to situations in which humans are exposed over relatively long periods to those agents. This is discussed in the next section.

4.4.2 Man

There is a good deal of controversy concerning possible behavioural effects of anticholinesterase agents on man. In particular there are differences in opinion about psychological and psychiatric sequelae of chronic exposure to such agents. The basic difficulty lies in the fact that most exposures are accidental so that there is usually only a rough idea or no idea at all of the degree of the exposure. Additionally, the individuals concerned, usually those who are exposed to insecticides are often potentially vulnerable to a variety of agents, not all of them anticholinesterases, over a considerable period of time.

The evidence which is available can be divided into three categories—the few reports of deliberate administration of anticholinesterase agents to man for clinical observation of the effects, case reports of individual patients, and surveys of relatively large populations who are at risk, usually those engaged in the manufacture or the spraying of insecticides.

An early report of the effects of administering dyflos to man was by Grob, Harvey, Langworthy and Lilienthal (1947). They gave 1 to 2 mg of the agent daily by intramuscular injection to 60 normal subjects. Symptoms apparently originating in the central nervous system included, in order of frequency, excessive dreaming, insomnia, jitteriness and restlessness, increased tension, emotional lability, subjective tremulousness, nightmares, headache, increased libido, giddiness, drowsiness, paresthesias, mental confusion, visual hallucinations, tremor and pains in the legs of sciatic distribution. Neostigmine, a quaternary ammonium salt which does not penetrate into the central nervous system, did not significantly affect these symptoms. Neither did d-tubocurarine. Atropine and barbituric acid derivatives reduced the symptoms to some extent.

Following this, 23 subjects were studied, 19 were normal while 4 had myasthenia gravis but normal EEG patterns. Dyflos, in the same dose as above, produced increased electrical activity of the brain voltage variations, higher frequencies, irregularities in rhythm and the sporadic appearance of slow waves similar to those seen in patients with grand mal epilepsy. These changes appeared after 2 to 7 days of dyflos administration and usually followed the onset of symptoms of the type listed above. When the drug was discontinued the symptoms disappeared within 1 to 4 days but the EEG changes persisted in diminishing degree for 8 to 42 days (average 29 days).

When dyflos was administered for periods of up to 3 days the onset of symptoms and EEG changes usually coincided with depression of erythrocyte cholinesterase levels to about 70 and 60 per cent of normal, respectively. No such correlation could be demonstrated when the drug was administered for longer periods or when drug administration ceased. No correlations at all could be detected between symptoms or EEG changes and plasma cholinesterase levels.

Atropine (1.2 mg intravenously) reversed the EEG changes in all subjects and atropine given by a variety of routes concomitant with the daily dyflos injection delayed the onset of symptoms and EEG changes. The latter only appeared after 3 or 4 weeks when the erythrocyte cholinesterase had fallen to an average of 33 per cent of normal. Central nervous system symptoms were only delayed in onset by a few days and were present in several subjects when the EEG was normal. In these subjects when the administration of atropine was terminated EEG abnormalities appeared in a few days. The EEG effects, as well as the central nervous system symptoms were unaffected by both neostigmine and d-tubocurarine.

Two of the subjects with upper motor neurone lesions showed increased spinal cord activity following dyflos. There were also spontaneous discharges within the spinal cord below the level of the lesions.

More recently Bowers, Goodman and Sim (1964) have reported the effects on 93 male volunteers of an organophosphate, code number EA-1701, the chemical nature of which was not revealed in the paper. Neither is the dose of agent given nor the precise site of application although it is stated that it "was applied by a micrometer syringe to various body sites" where it was presumably absorbed percutaneously. Erythrocyte, plasma and whole blood cholinesterase levels were determined at intervals and it transpired that symptoms correlated well with depressions in erythrocyte or whole blood, but not plasma, cholinesterase. These depressions were usually maximal at three to eight hours after application of the agent. A check-list was used to systematise clinical observations which were mainly directed towards psychological effects of the agent, the only physical effects noted being nausea and vomiting. When both mental and physical effects were seen in the same subject the former usually preceded the latter. Initially, subjects reported vague feelings of fatigue and difficulties in concentrations. Later "jitteriness" or "tenseness inside" were reported and these feelings persisted at a low intensity for the 24 hour period of observation. During this time many subjects experienced difficulties in maintaining a train of thought and became depressed, irritable and listless. There were minor difficulties with orientation, mainly slow responses to date, time and location. However, all the subjects retained insight and were aware of the decrements in their performance and there were no illogical or inappropriate trends in thought or speech.

In general, psychological effects did not appear until whole blood cholinesterase levels fell to at least 40 per cent of control, although some subjects reported post-experimental dreams at cholinesterase levels of 41 to 80 per cent of control. About half the subjects with cholinesterase levels between 10 and 40 per cent of controls showed psychological effects as assessed by clinical observation. The rating of this group on Clyde Mood scores (a card-sorting test) was significantly different than their pre-drug rating. On the basis of this test the subjects were considered to be less friendly, clear-thinking and energetic but more jittery and depressed.

In most affected subjects the psychological effects were most marked in the period 8 to 10 hours after drug administration but there was a waxing and waning of these effects over a 24 hour period. Usually, there were unpleasant dreams during the first night after the drug but by the next morning in most cases all effects had disappeared. In those individuals who experienced marked nausea and vomiting, treatment with atropine was instituted and the impression was that this was beneficial for both the physical and the psychological effects.

These two studies, by Grob *et al.* (1947) and Bowers, Goodman and Sim (1964) serve to show the kind of clinical picture, particularly from the behavioural viewpoint, which anticholinesterase agents produce. The two accounts seem to agree well and in both studies, irrespective of whether a single dose or multiple doses of the agents were given, the effects described were, on the evidence given, entirely reversible.

There are also a number of papers which describe effects on individuals. Two examples will serve to show the kind of information which can be obtained in this way. Dille and Smith (1963) studied two pilots who had been spraying insecticides. The first one had received psychotherapy for eight months for an anxiety reaction with tension headaches. Just before and during this period of treatment he was engaged in spraying organophosphate insecticides. Little improvement occurred in his mental condition and he was admitted to hospital where the initial diagnosis was of a neurotic reaction with obsessive compulsive, phobic and depressive symptoms. Electro-shock therapy was given and the patient was discharged after two months on iproniazid and pyridoxine. The final diagnosis was psychotic depressive reaction. The patient claimed to have done no commercial flying during the two years after his discharge from hospital but during this time he passed two annual physical examinations for the issue of a second class medical certificate enabling him to fly. During the third year he flew 600 hours and at the end of this period his medical record including a current psychiatric evaluation was reviewed by the Civil Air Surgeons Medical Review Board and he was issued with a medical certificate to fly.

The second pilot had been spraying a variety of insecticides including methyl parathion, DDT, toxaphene, eldrin and dieldrin more or less regularly for a period of about ten years. During that time he had experienced repeated bouts of nausea and diarrhoea and early in this period he had been advised to stop spraying because of liver and kidney damage. The most dramatic episode seems to have been during the eighth year of this ten year period when the pilot was accidentally saturated with a mixture of parathion, toxaphene and Dipterex. He became ill and was admitted to hospital with nausea, abdominal cramps, dizziness, tightness of the chest, weakness of hand and legs and a rash on his wrists. His blood cholinesterase fell to 12 per cent of normal, and there was evidence of liver disease. After a lay-off of some ten months he resumed aerial spraying but experienced nausea, weakness and dizziness and an examination revealed an elevated white cell count which was attributed to tonsillitis; after a tonsillectomy he again returned to work but experienced the same symptoms. Soon he visited the Civil Aeromedical Research Institute seeking help for recurrent bouts of acute anxiety which he attributed to long exposure to organophosphate insecticides. His EEG showed certain abnormalities at this time. Unfortunately there is no record of whether this second pilot eventually recovered.

Another paper, concerned with three cases, is by Urban and Sando (1965). The first case was a man who swallowed possibly 60 ml of malathion and died 19 hours later despite treatment with atropine and 2-PAM. The second case was a women who accidentally drank parathion. She was also treated with atropine and 2-PAM and after four days was sufficiently recovered to carry on simple conversations. She then showed a rapid deterioration with screaming and restlessness and apparent unawareness of her surroundings. This continued for about three weeks but then a gradual improvement began and by 27 weeks after poisoning she was managing her domestic duties but was mentally slow and her gait was not quite steady. There were severe ophthalmological disorders and a generalised abnormality in the EEG though the latter showed improvement from the recording taken 10 weeks after poisoning. Blood cholinesterase levels were monitored in this patient. Plasma enzyme levels had returned to normal within 14 days and erythrocyte levels within 10 weeks. The third case was a man who attempted suicide by drinking malathion. He was given atropine and 2-PAM and three days after admission to hospital he was transferred to a mental hospital for treatment of his severe depression which, it must be assumed, was present before his suicide attempt since the authors stated categorically that there were no sequelae attributable to malathion.

These few case histories serve to indicate the difficulties inherent in attempting to draw general conclusions from isolated examples. The two pilots had been exposed to, or were potentially at risk from, a variety of insecticides, not all organophosphates. However, it seems not unreasonable to conclude that many of the effects they experienced were due to cholinesterase inhibition. In one there seemed to be no residual effects after a period since he was passed fit to fly while in the other the reported follow-up period was not sufficient to know whether recovery occurred. The other three cases were different in that they received massive single doses rather than cumulative doses, and of these one died despite treatment, one recovered, presumably due to treatment and one showed clear residual effects which, nevertheless, showed progressive lessening over the period of the study.

This variability in response points to the need for studies of larger numbers of subjects and this is where the surveys of relatively large populations at risk provide useful information.

A population of intermediate size (16 cases) was reported on by Gershon and Shaw (1961). The individuals concerned were 3 scientific officers who were studying the efficacy of insecticide sprays, 8 greenhouse workers and 5 farm workers. Their periods of exposure ranged from 1½ to 10 years. It was claimed that 7 showed depressive reactions, 5 schizophrenic reactions and 1 a fugue. All 16 had disturbances of memory and concentration, 11 had acute effects with the expected signs and symptoms of anticholinesterase poisoning. Follow up of the cases showed that the effects persisted for at least six months after cessation of exposure but almost all reverted to normal in twelve months.

As a result of these findings a field survey was carried out by asking local physicians in fruit growing districts whether mental ill-health had increased and whether country dwellers were more affected than town dwellers. Most of the 16 physicians consulted believed that mental ill-health had not increased but 3 thought that depression was commoner in orchard workers.

Gershon and Shaw admitted that their results had not been analysed statistically and were drawn from the community at large and that the objection could be made that their results were due to chance, that is, the exposure to organophosphates and mental change are not correlated. It was on these and other grounds that their paper was criticized by three other authors.

Barnes (1961) noted that no figures for blood cholinesterase activity of any of the 16 cases were quoted by Gershon and Shaw and that no facts were given about the incidents in which exposure to organophosphates was said to have occurred. He compared the findings of Gershon and Shaw with those of Holmes and Gaon (1956) which are discussed below and noted that they were very considerably at variance. He also mentioned that Gershon and Shaw claimed to have diagnosed psychiatric disorders in 8 out of 10 greenhouse workers and made the point that an incidence of 80 per cent of disorders of this type would not have been likely to have escaped the attention of others previously.

In another criticism of Gershon and Shaw's paper, Bidstrup (1961) noted that hundreds of cases of acute poisoning with organophosphorus insecticides had been recorded in countries other than Australia but that in none of them had psychiatric symptoms of more than a transient nature been reported. He made the plea that any conclusions about possible psychiatric sequelae of chronic exposure to organophosphorus compounds should be drawn from controlled observations rather than from impressions.

Finally, a detailed criticism of the paper of Gershon and Shaw was made by Golz (1961). His criticisms were more or less along the same lines as those of Barnes and Bidstrup.

The paper by Holmes and Gaon (1956) referred to by Barnes (1961) is concerned with a review of over 600 cases of exposure to anticholinesterase agents. These included insecticides, particularly parathion or TEPP, and also nerve gas. The incidence of symptoms of various kinds in these cases is reviewed. Symptoms attributable to central nervous system effects of the drugs included headache, dreams, fatiguability, nervousness and irritability, increased perspiration, dizziness, tremor and twitch, paresthesia and cold. These occurred relatively infrequently in the milder cases of poisoning but their incidence rose in the more severe cases. An attempt was made to relate the incidence of symptoms to the degree of inhibition of erythrocyte cholinesterase in a group of 449 cases. Of these, 169 showed 0 to 10 per cent reduction in cholinesterase activity and in this group, for example, 43 per cent reported headache, 33 per cent dreams and poor sleep, 24 per cent nervousness and irritability and 11 per cent dizziness. In contrast there were 16 cases with erythrocyte cholinesterase depletions of over 60 per cent and the corresponding incidence of the same signs or symptoms was, headache 69 per cent, dreams and poor sleep 75 per cent, nervousness and irritability 37 per cent and dizziness 31 per cent.

The duration of these signs and symptoms is not given in the paper but the implication is that they were completely reversible since specific mention is made of another group who have had multiple as opposed to single exposures and who commonly complained of persistent symptoms. As well as muscle aches, pains, numbness and weakness of limbs these symptoms include irritability, forgetfulness, confusion in thinking and the need for marked concentration in performing certain tasks like driving a car or building a radio set. A review was made of 37 cases who had received 3 or more exposures, for the signs and symptoms exhibited on their most recent exposure. It was noted that the duration of symptoms tended to correlate with the degree of reduction in erythrocyte cholinesterase. In this group there was a fairly high incidence of irritability and forgetfulness.

EEGs were done on a number of the cases. Focal and paroxysmal changes were observed shortly after exposure but these had usually disappeared on subsequent examination. In one severe case of parathion poisoning EEG abnormalities persisted to the time of writing which was one year after the exposure.

A later survey of EEG, psychological and neurological alterations in a group of individuals engaged in the manufacture of organophosphorus compounds was made by Metcalf and Holmes (1969). These were studied from 1965 to 1969 although many had been employed in the same capacity since 1950.

The results of psychological tests indicated that the dysfunctions most clearly seen in the exposed group were disturbed memory, and difficulty in maintaining alertness and focusing attention. When the paper was written it was not possible to ascertain whether there were any precise relationships between these dysfunctions and work history, cholinesterase levels or severity, type and duration of exposure.

Approximately equal numbers of highly exposed and minimally exposed workers were given neurological examinations. No differences between the two groups were detected in "hard" neurological signs like sensory or motor deficits but there was an impression that exposed men showed more "soft" neurological signs like minor coordination deficits and oculomotor imbalance.

Analysis of the EEG records of exposed men showed no increase in the incidence of "hard" EEG abnormalities like spike activity or focal slowing. There was, however, a high incidence of theta activity, that is, 4-6 Hz waves, occurring during light drowsiness in short episodes of 2–4 seconds duration. Although this is a minimal disturbance of the EEG the interesting and perhaps significant point is that it mirrors the more severe EEG disturbances which are seen after acute exposure. It was only followed up at length, for one year, in one man in whom it showed a gradual decline.

The authors speculated that the central effects observed were causally related to actions of the organophosphates on deep midbrain pontine centres, rich in acetylcholinesterase.

Another survey of a large population was by Sumerford, Hayes, Johnston, Walker and Spillane (1953) who studied 258 people from the area of Wenatchee, Washington, U.S.A. which is the centre of extensive orchards where organophosphorus insecticides were in common use. Included among these people were some who mixed or applied the insecticides, that is those who had extensive occupational exposure. A second group included miscellaneous workers who were known to have definite but irregular and often minimal exposures. There were two groups of residents living near and far from the orchards respectively. Finally, there was a group who lived outside the area completely. These latter had no opportunity for exposure.

Of these 258 individuals, 246 contributed at least one blood sample and 223 were studied in enough detail to have their symptomatology considered in the report. The blood cholinesterase values in the "pre-exposure" period showed some variability but at least some persons who were known to have definite and consistent exposure to organophosphate insecticides showed marked depletion of these values during the spraying season. This was particularly marked in mixing-plant personnel, less marked but still significantly so in insecticide applicators and very slight and of doubtful significance in the miscellaneous group of workers. The maximum depletions occurred during August; by November there was complete recovery.

In reviewing the symptomatology data fifty nine symptoms were recorded, at least once. Some were associated with established diagnoses in a few of the subjects. Further analysis used twenty three symptoms selected mainly on the basis of their known or suspected association with intoxication. The percentages of observed illnesses involving at least one of these symptoms showed a random distribution in regard to exposure periods and occupational groups. However, when these illnesses which involved either miosis (a sign clearly associated with exposure) or at least three of the other selected symptoms were tabulated it was apparent that no such illness was seen during the pre-exposure or post-recovery periods and only once during the recovery period. The most commonly reported symptoms during the exposure period were headache, nausea, weakness or fatigue and pain or tightness in the chest. Behavioural effects, grouped as nervousness, drowsiness or insomnia, were relatively uncommon. In the thirty eight illnesses during the exposure period which reached the arbitrarily set criterion of involving miosis or at least three of the other selected symptoms, the symptoms mentioned above occurred the following number of times

Headache	29
Nausea	21
Weakness or fatigue	19
Pain or tightness in the chest	17
Nervousness, drowsiness or insomnia	9

This survey was continued in the 1952 and 1953 seasons and the results were given by Hayes, Dixon, Batchelor and Upholt (1957). The procedure was essentially the same as that used by Sumerford *et al.* (1953) in the 1951 survey and which is outlined above. The conclusions of the 1952–53 surveys were essentially similar to those from the earlier survey and confirmed relationships between blood cholinesterase depletions, exposure and illness. The later survey also showed clearly that illness resembling mild poisoning was not more common in persons living in this area who had no occupational or gross accidental exposure than in persons living in a non-agricultural area where insecticides were not used.

A similar survey involving 441 apple growers was reported by Davignon, St-Pierre, Charest and Tourangeau (1965). 170 persons living in the same environment together with 162 other persons having no contact with insecticides were used as controls. The apple growers had used a variety of pesticides including organophosphates, chlorinated hydrocarbons, naphthaquinone, chlorinated thiohydrocarbons, carbamates, amines, organic mercurials and inorganic salts. This, of course, makes it difficult to assess the causative factors involved in any deleterious effects. The latter were mainly neurological with an increased incidence of miosis, weakening or loss of reflexes, tremors and disturbances of equilibrium in the combined group of contacts and those living in the environment of the apple orchards as compared with the control groups.

Durham, Wolfe and Quinby (1965) made a particular study of mental alertness on groups of people with varying degrees of exposure to organophosphates and compared the results with findings from individuals with no known exposures. Again, the subjects were involved in various ways in the mixing or application of insecticides. Pencil and paper tests, essentially self-paced vigilance tests, were used and also an apparatus designed to measure complex reaction times.

In the first part of the survey carried out in 1960 only the two pencil and paper tests were used. The exposure group was divided into a number of sub-groups ranked in order of the amount of exposure but on no parameter of either test was there any indication that variation in effect correlated with difference in amount of exposure.

Comparisons between the performance of the control group and the exposed group as a whole were complicated by the fact that the control group performed less well in some aspects of the tests during the time of year corresponding to the "exposure" period for the exposed group. It seemed possible, therefore, that decrements in performance seen in the exposed group during this time may have been related to some factor other than exposure to insecticides. However, in one specific parameter the exposure group as a whole scored lower during the exposure period than during the non-exposure period whereas the control group scored better during the latter period.

In the 1961–62 survey with regard to every parameter studied in the reaction time test both the exposed group and the control group scored as well or better during non-exposure than exposure periods.

The general conclusions drawn by the authors were that no measurable difference in performance in the self-paced vigilance tests correlated with exposure to organophosphates and that the results of the complex reaction time test did not indicate that exposure to the agents had any significant effect on mental alertness.

However, the test results did seem to indicate that where amounts of the substances great enough to cause clinical signs or symptoms of mental illness were absorbed then mental alertness was affected. None of the individuals in this survey showed mental effects without clinical signs and symptoms.

The final paper to be considered here is by Stoller, Krupinski, Christophers and Blanks (1965) who, as a result of the paper by Gershon and Shaw (1961) discussed above, instituted an epidemiological study in Australia to determine whether areas of high organophosphate usage had a higher proportion of hospital admission for psychiatric disorders than did low-usage areas. The sales of organophosphates in the non-metropolitan areas of Victoria during 1963 were analysed. Six areas were covered and the two with the highest proportion of sales selected for further study. These were the Goulburn Valley and the Murray Valley. Neither of these areas showed a higher incidence of psychiatric disorders per 100,000 of the male population than non-metropolitan areas chosen as controls. It was found that the Goulburn Valley had a significantly higher proportion of schizophrenic breakdowns than did the Murray Valley. However all the cases from the Goulburn Valley were followed up and it was found that none had worked on a fruit farm and none had been exposed to organophosphorus insecticides.

These examples have been quoted at some length and in some detail to indicate the conflicting evidence which has been produced concerning long term behavioural effects of anticholinesterases. Bearing in mind this confusion it is impossible to make any unequivocal statements but it is perhaps reasonable to draw some tentative conclusions. It should be emphasised that these relate specifically to psychiatric or psychological sequelae and not to neurological sequelae of the type discussed in Section 4.3.3 of this Chapter. There is no doubt that the latter can occur following absorption of even single doses of some types of agent.

There are adequate accounts of mental effects following acute or chronic exposures but on present evidence it seems that they only occur when there is appreciable inhibition of blood cholinesterase and that they are often accompanied by typical peripheral signs of poisoning. These effects seem to be completely reversible. The more vexed question concerns the possible long-term or even permanent effects occurring in individuals with multiple exposures to anticholinesterase agents. The concensus of opinion seems to be that behavioural changes of a relatively minor nature, characterised by irritability, difficulty in concentration and fatigue, do occur in some such individuals but in most the changes progressively decrease when exposure ceases. Unfortunately there is insufficient evidence to show whether recovery has been complete in all cases. It should be emphasised, however, that most of the individuals concerned have had considerable exposures and considerable depletions in their blood cholinesterase.

It is difficult to quarrel with a comment by West (1968) who states "The literature on potential, suspected and established sequelae of phosphate ester poisoning does not confirm the often repeated statement that clinical recovery from nonfatal poisoning is always complete in a few days. On the other hand, no easily recognised, serious or permanent sequelae have been sufficiently frequent to make themselves obvious".

References

Ågren, G. and Ramachandran, B. V. (1963). The effect of pyridinium aldoximes and atropine on the incorporation of DFP^{32} in rat liver cell fractions. *Acta Physiol. Scand.*, **60**, 95–102

Aldridge, W. N. and Barnes, J. M. (1961). Neurotoxic and biochemical properties of some triaryl phosphates. *Biochem. Pharmac.*, **6**, 177–188

Aldridge, W. N. and Barnes, J. M. (1966a). Further observations on the neurotoxicity of organophosphorus compounds. *Biochem. Pharmac.*, **15**, 541–548

Aldridge, W. N. and Barnes, J. M. (1966b). Esterases and neurotoxicity of some organophosphorus compounds. *Biochem. Pharmac.*, **15**, 549–554

Aldridge, W. N., Barnes, J. M. and Johnson, M. K. (1969). Studies on delayed neurotoxicity produced by organophosphorus compounds. *Ann. N.Y. Acad. Sci.*, **140**, 314–322

Arterberry, J. D., Boniface, R. W., Nash, E. W. and Quinby, G. E. (1962). Potentiation of phosphorus insecticides by phenothiazine derivatives. *Science*. **182**, 848–850

Askew, B. M. (1956). Oximes and hydroxamic acids as antidotes in anti-cholinesterase poisoning. *Br. J. Pharmac. Chemother.*, **11**, 417–423

Askew, B. M. (1957). Oximes and atropine in sarin poisoning. *Br. J. Pharmac. Chemother.*, **12**, 340–343

Axelrod, J. (1956). Possible mechanism of tolerance to narcotic drugs. *Science*, **124**, 263

Banks, A. (1957). Effects of an anticholinesterase on problem solving behaviour in the white rat. PhD Thesis, University of London

Barlow, R. B. (1964). *Introduction to Chemical Pharmacology*. Methuen & Co. Ltd., London

Barnes, J. M. (1961). Psychiatric sequelae of chronic exposure to organophosphorus insecticides. *Lancet*, **2**, 102–103

Barnes, J. M. and Denz, F. A. (1954). The reaction of rats to diets containing octamethyl pyrophosphoramide (Schraden) and O,O-diethyl-S-ethylmercaptoethanol thiophosphate (Systox). *Br. J. Indust. Med.*, **11**, 11–19

Barstad, J. A. B. (1956). The effect of d-tubocurarine on the neuromuscular blocks caused by diisopropylfluorophosphate and acetylcholine. *Arch. int. Pharmacodyn.*, **107**, 4–20

Barstad, J. A. B. (1960). Cholinesterase inhibition and the effect of anticholinesterases on indirectly evoked single and tetanic muscle contractions in the phrenic nerve-diaphragm preparation from the rat. *Arch. int. Pharmacodyn.*, **128**, 143–168

Bay, E. and Adams, N. L. (1968). Evidence for the existence of nicotine sensitive receptors in the medullary respiratory centres of cats. *Pharmacologist*, **10**, 174

Bebbington, A. and Brimblecombe, R. W. (1965). Muscarinic receptors in the peripheral and central nervous systems. *Adv. Drug. Res.*, **2**, 143–172

Berry, W. K. and Davies, D. R. (1970). The use of carbamates and atropine in the protection of animals against poisoning by 1,2,2-trimethylpropyl methylphosphonofluoridate. *Biochem. Pharmac.*, **19**, 927–934

Berry, W. K., Davies, D. R. and Rutland, J. P. (1966). Problems in the treatment with oximes and atropine of rats poisoned by organophosphates. *Biochem. Pharmac.*, **15**, 1259–1266

Bethe, K., Erdmann, W. D., Lendle, L. and Schmidt, G. (1957). Spezifische Antidot Behandlung bei protrahierter Vergiftung mit Alkylphosphaten. (Paraoxon, Parathion, DFP) und Eserin an Meerschweinchen. *Naugyn-Schmiedeberg's Arch. exp. Path Pharmak.*, **231**, 3–22

Bidstrup, P. L. (1961). Psychiatric sequelae of chronic exposure to organophosphorus insecticides. *Lancet*, **2**, 102–103

Bignami, G. and Gatti, G. L. (1967). Neurotoxicity of anticholinesterase agents. Antagonistic action of various centrally acting drugs. *Proceedings of the European Society for the Study of Drug Toxicity*. Vol. VIII. *Neurotoxicity of Drugs*. Excerpta Medica Foundation, Amsterdam, pp 93–106

Blaschko, H., Bulbring, E. and Chou, T. C. (1949). Tubocurarine antagonism and inhibition of cholinesterases. *Br. J. Pharmac. Chemother.*, **4**, 29–32

Bowers, M. B., Goodman, E. and Sim, V. M. (1964). Some behavioural changes in man following anticholinesterase administration. *J. Nerv. Ment. Dis.*, **138**, 383–389

Bradley, P. B. and Elkes, J. (1953). The effect of atropine, hyoscyamine, physostigmine and neostigmine on the electrical activity of the brain of the conscious cat. *J. Physiol. Lond.*, **120**, 14–15P

Bradley, P. B. and Elkes, J. (1957). The effect of some drugs on the electrical activity of the brain. *Brain*, **80**, 77–117

Brimblecombe, R. W. (1963). Effects of psychotropic drugs on open-field behaviour in rats. *Psychopharmacologia, Berl.*, **4**, 139–147

Brimblecombe, R. W. (1970). The use of animal tests to predict behavioural effects of chemicals on man. *Chemical Influences on Behaviour*. (eds. R. Porter and J. Birch) J. and A. Churchill, London, pp 6–24

Brimblecombe, R. W. and Everett, S. D. (1969). Interaction between sarin and the cholinergic antagonist PMCG in fast and slow skeletal muscle of the cat. *J. Physiol. Lond.*, **203**, 19–20P

Brimblecombe, R. W. and Everett, S. D. (1970). Actions of sarin on fast-twitch and slow-twitch skeletal muscles of the cat and protective action by anticholinergic drugs. *Br. J. Pharmac.*, **40**, 57–67

Brimblecombe, R. W., Green, D. M., Stratton, J. A. and Thompson, P.B.J. (1970). The protective actions of some anticholinergic drugs in sarin poisoning. *Br. J. Pharmac.*, **39**, 822–830

Brimblecombe, R. W., Coult, D. B., Deane, C. C. and Parkes, D. C.

(1971). Biochemical and behavioural effects of some halo-substituted vinyl phosphorus esters. *Biochem. Pharmac.*, **20**, 1733–1737

Carson, V. G., Jenden, D. J. and Russell, R. W. (1973). Changes in peripheral cholinergic systems following development of tolerance to the anticholinesterase diisopropylfluorophosphate. *Toxicol. appl. Pharmac.*, **26**, 39–48

Casida, J. E., Baron, R. L., Eto, M. and Engel, J. L. (1963). Potentiation and neurotoxicity induced by certain organophosphates. *Biochem. Pharmac.*, **12**, 73–83

Cavanagh, J. B. (1964). Peripheral nerve changes in orthocresylphosphate poisoning in the cat. *J. Path. Bact.*, **87**, 365–383

Childs, A. F., Davies, D. R., Green, A. L. and Rutland, J. P. (1955). The reactivation by oximes and hydroxamic acids of cholinesterase inhibited by organophosphorus compounds. *Br. J. Pharmac. Chemother.*, **10**, 462–465

Coleman, I. W., Little, P. E. and Bannard, R. A. B. (1962a). Cholinolytics in the treatment of anticholinesterase poisoning. I. The effectiveness of certain cholinolytics in combination with oxime for treatment of sarin poisoning. *Canad. J. Biochem. Physiol.*, **40**, 815–826

Coleman, I. W., Little, P. E. and Bannard, R. A. B. (1962b). Cholinolytics in the treatment of anticholinesterase poisoning. II. Treatment of sarin poisoning with an oxime and various combinations of cholinolytic compounds. *Canad. J. Biochem. Physiol.*, **40**, 827–834.

Coleman, I. W., Little, P. E. and Bannard, R. A. B. (1963). Cholinolytics in the treatment of anticholinesterase poisoning. III. The effects of quaternisation and alteration of anionic function on the potency of cholinolytics in the treatment of sarin poisoning. *Canad. J. Biochem. Physiol.*, **41**, 2479–2491.

Cox, B. and Potkonjak, D. (1969a). The relationship between tremor and change in brain acetylcholine concentration produced by injection of oxotremorine in the rat. *Br. J. Pharmac.*, **35**, 295–303

Cox, B. and Potkonjak, D. (1969b). The effects of atropine and dyflos on tremor and increase in whole brain acetylcholine produced by injection of oxotremorine in the rat. *Br. J. Pharmac.*, **35**, 521–529

Cullumbine, H. (1963). Actions at autonomic effector sites. *Handbuch. der Experimentellen Pharmakologie. Cholinesterases and Anticholinesterases.* (ed. G. B. Koelle) Springer-Verlag, Berlin, pp 505–529

Dahlbom, R., Edlund, T., Ekstrand, T. and Katz, A. (1952). The ability of some phenothiazine derivatives to inhibit nicotine-induced tremors. *Arch. int. Pharmacodyn.*, **90**, 241–250

Dahlbom, R., Diamant, H., Edlund, T., Ekstrand, T. and Holmstedt, B. (1953). Protective effect of phenothiazine derivatives against poisoning by the irreversible cholinesterase inhibitor dimethylamidoethoxyphosphoryl cyanide (tabun). *Acta pharmacol.* (*Kbh.*), **9**, 163–167

Davies, D. R. (1963). Neurotoxicity of organophosphorus compounds. *Handbuch der Experimentellen Pharmakologie. Cholinesterases and Anticholinesterases.* (ed. G. B. Koelle) Springer-Verlag, Berlin, pp 860 882

Davies, D. R. and Green, A. L. (1955). *The Physical Chemistry of Enzymes.* Disc. Faraday Soc. No. 20, Aberdeen University Press Ltd., p 269

Davies, D. R., Holland, P. and Rumens, M. J. (1960). The relationship between the chemical structure and neurotoxicity of alkyl organophosphorus compounds. *Br. J. Pharmac. Chemother.*, **15**, 271–278

Davignon, L. F., St-Pierre, J., Charest, G. and Tourangeau, F. J. (1965). A study of the chronic effects of insecticides in man. *Canad. Med. Ass. J.*, **92**, 597–602

Despierre, F. and Martin, L. (1958). Anticholinesteragiques. Protection *in vitro* par les ammoniums quaternaires. *C.R. Acad. Sci.*, **246**, 183–186

Deutsch, J. A. (1969). The physiological basis of memory. *Ann. Rev. Psychol.*, **20**, 85–104

Dille, J. R. and Smith, P. W. (1963). *Central nervous system effects of chronic exposure to organophosphate insecticides.* Federal Aviation Agency, Oklahoma City, Oklahoma

Distefano, V., Hurwitz, L., Neuman, W. F. and Hodge, H. C. (1951). Curamine (Nikethamide) as adjuvant to atropine in treatment of poisoning by EPN. *Proc. Soc. exp. Biol.* N.Y., **78**, 712–713

Dubois, K. P., Doull, J., Salerno, P. R. and Coon, J. M. (1949). Studies on the toxicity and mechanism of action of p-nitrophenyl diethyl thiophosphate (parathion). *J. Pharmac. exp. Ther.*, **95**, 79–91

Durham, W. F., Wolfe, H. R. and Quinby, G. E. (1965). Organophosphorus insecticides and mental alertness. *Arch. environ. Hlth.*, **10**, 55–60

Earl, C. J. and Thompson, R. H. S. (1952a). The inhibitory action of tri-ortho-cresyl phosphate on cholinesterases. *Br. J. Pharmac. Chemother.*, 7, 261–269

Earl, C. J. and Thompson, R. H. S. (1952b). Cholinesterase levels in the nervous system in tri-ortho-cresyl phosphate poisoning. *Br. J. Pharmac. Chemother.*, 7, 685–692

Edery, H. and Schatzberg-Porath, G. (1958). Pyridine-2-aldoxime methiodide and diacetyl monoxide against organophosphorus poisoning. *Science,* **128**, 1137–1138

Engelhard, N. and Erdmann, W. D. (1964). Beziehungen zwischen chemischer struktur und Cholinesterase reaktivierender wirksamkeit bei einer Reihe neuer bis-quartärer pyridin-4-aldoxime. *Arzneim.-forsch.*, **14**, 870–875

Feldberg, W. and Sherwood, S. L. (1954). Behaviour of cats after intraventricular injections of eserine and DFP. *J. Physiol. Lond.*, **125**, 488–500

Fleisher, J. H., Corrigan, J. P. and Howard, J. W. (1958). Potentiation of the response of frog rectus muscle to acetylcholine by isopropyl methyl phosphonofluoridate and its modification by pyridine-2-aldoxime methiodide. *Br. J. Pharmac. Chemother.*, **13**, 291–295

Fleisher, J. H., Moen, T. H. and Ellingson, N. R. (1965). Effects of 2-PAM and TMB-4 on neuromuscular transmission. *J. Pharmac. exp. Ther.*, **149**, 311–319

Fleisher, J. H., Harris, L. W., Miller, G. R., Thomas, N. C. and Cliff, W. J. (1970). Antagonism of sarin poisoning in rats and guinea pigs by atropine, oximes and mecamylamine. *Toxicol. Appl. Pharmac.*, **16**, 40–47

Frada, G. and Gucciardi, G. (1957). Influenza della chlorpromazina sull' intossicazione da esteri forforici. *Med. Lav.*, **48**, 301–306

Freedman, A. M. and Himwich, H. E. (1949). DFP: site of injection and variation in response. *Amer. J. Physiol.*, **156**, 125–128

Funderburk, W. H. and Case, T. J. (1947). Effect of parasympathetic drugs on the conditioned response. *J. Neurophysiol.*, **10**, 179–190

Gaines, T. (1962). Poisoning by organic phosphorus pesticides potentiated by phenothiazine derivatives. *Science*, **138**, 1260–1261

Gershon, S. and Shaw, F. H. (1961). Psychiatric sequelae of chronic exposure to organophosphorus insecticides. *Lancet*, **1**, 1371–1374

Glow, P. H. and Rose, S. (1965). Effects of reduced acetylcholinesterase levels on extinction of a conditioned response. *Nature, Lond.*, **206**, 475–477

Goldberg, M. E., Johnson, H. E., Kraak, J. B. and Smyth, H. F. (1963). Psychopharmacological effects of reversible cholinesterase inhibition induced by N-methyl-3-isopropyl phenyl carbamate. (Compound 10854). *J. Pharmac. exp. Ther.*, **141**, 244–252

Golz, H. H. (1961). Psychiatric sequelae of chronic exposure to organophosphorus insecticides. *Lancet*, **2**, 369–370

Goodman, L. S. and Gilman, A. (1970). *The Pharmacological Basis of Therapeutics*. (eds. L. S. Goodman and A. Gilman). The Macmillan Co., New York, pp 447–456

Goyer, G. R. (1968). Action of pyridine-2-aldoxime methochloride (PAM) on neuromuscular transmission *in vitro* and *in vivo*. *Canad. J. Physiol. Pharmac.*, **46**, 757–764

Goyer, G. R. (1969). Action of pyridine-2-aldoxime methochloride (PAM) on the rectus abdominus muscle of the frog. *Canad. J. Physiol. Pharmac.*, **47**, 1009–1014

Grob, D. (1963). Anticholinesterase intoxication in man and its treatment. *Handbuch der Experimentellen Pharmakologie. Cholinesterases and Anticholinesterases.* (ed. G. B. Koelle) Springer-Verlag, Berlin, pp 989–1027

Grob, D. and Johns, R. J. (1958). Use of oximes in the treatment of intoxication by anticholinesterase compounds in normal subjects. *Amer. J. Med.,* **24**, 497–511

Grob, D., Harvey, A. M., Langworthy, O. R. and Lilienthal, J. L. (1947). The administration of diisopropyl fluorophosphate (DFP) to man. III. Effect on the central nervous system with special reference to the electrical activity of the brain. *Bull. Johns Hopkins Hosp.,* **81**, 257–266

Harwood, C. T. (1954). Cholinesterase activity and electroencephalograms during circling induced by the intracarotid injection of di-isopropyl fluorophosphate (DFP). *Amer. J. Physiol.,* **177**, 171–174

Hayes, W. J., Dixon, E. M., Batchelor, G. S. and Upholt, W. M. (1957). Exposure to organic phosphorus sprays and occurrence of selected symptoms. *Public Health Rep.,* **72**, 787–794

Head, H. (1889). On the regulation of respiration. *J. Physiol., Lond.,* **10**, 1–40

Hobbiger, F. (1957a). The protective action of pyridine-2-aldoxime methiodide in organophosphate poisoning. *Biochem. J.,* **66**, 7P

Hobbiger, F. (1957b). Protection against the lethal effects of organophosphates by pyridine-2-aldoxime methiodide. *Br. J. Pharmac. Chemother.,* **12**, 438–446

Hobbiger, F. (1963). Reactivation of phosphorylated acetylcholinesterase. *Handbuch der Experimentellen Pharmakologie. Cholinesterases and Anticholinesterases.* (ed. G. B. Koelle) Springer-Verlag, Berlin, pp 921–988

Hobbiger, F. and Sadler, P. W. (1959). Protection against lethal organophosphate poisoning by quaternary pyridine aldoximes. *Br. J. Pharmac. Chemother.,* **14**, 192–201

Hobbiger, F., O'Sullivan, D. G. and Sadler, P. W. (1958). New potent reactivators of acetylcholinesterase inhibited by tetraethylpyrophosphate. *Nature, Lond.,* **182**, 1498–1499

Holmes, J. H. and Gaon, M. D. (1956). Observations on acute and multiple exposure to anticholinesterase agents. *Trans. Amer. clin. clim. Ass.,* **68**, 86–101

Holmes, R. and Robins, E. L. (1955). The reversal by oximes of neuromuscular block produced by anticholinesterases. *Br. J. Pharmac. Chemother.,* **10**, 490–495

Holmstedt, B. (1959). Pharmacology of organophosphorus cholinesterase inhibitors. *Pharmac. Rev.*, **11**, 567–688

Holmstedt, B. (1963). Structure-activity relationships of the organophosphorus anticholinesterase agents. *Handbuch der Experimentellen Pharmakologie. Cholinesterases and Anticholinesterase Agents.* (ed. G. B. Koelle) Springer-Verlag, Berlin, pp 428–485

Jovic, R. and Boskovic, B. (1970). Antidotal action of pyridinium oximes in poisoning by O,O-diethyl-S-/2-(N-methyl-N-phenylamino) ethyl/ thiophosphate methylsulfomethylate (GT-45) and its two new analogs. *Toxicol. Appl. Pharmac.*, **16**, 194–200

Kagan, I. S. (1962). The effect of thiophosphates on conditioned reflexes and the cholinesterase activities of cats. *J. high. nerv. Act.*, **12**, 530–540

Karczmar, A. G. (1969). Is the central cholinergic nervous system overexploited? *Fed. Proc.*, **28**, 147–157

Karczmar, A. G. and Long, J. P. (1958). Relationship between peripheral cholinolytic potency and tetraethylpyrophosphate antagonism of a series of atropine substitutes. *J. Pharmac. exp. Ther.*, **123**, 230–237

Kleinwachter, L. (1864). Beobachtung über die Wirkung des Calabar—Extracts gegen Atropin-Vergiftung. *Berl. Klin. Wschr.*, **1**, 369–371

Koelle, G. B. (1946). Protection of cholinesterase against irreversible inactivation by DFP *in vitro. J. Pharmac. exp. Ther.*, **88**, 323–327

Koelle, G. B. (1970). Anticholinesterase agents. *The Pharmacological Basis of Therapeutics.* 4th edition (eds. L. S. Goodman and A. Gilman) Macmillan, New York, pp 442–465

Koster, R. (1946). Synergisms and antagonisms between physostigmine and diisopropyl fluorophosphate in cats. *J. Pharmac. exp. Ther.*, **88**, 39–46

Krech, D., Rosenzweig, M. R. and Bennet, E. L. (1956). Dimensions of discrimination and level of cholinesterase activity in the cerebral cortex of the rat. *J. Comp. Physiol. Psychol.*, **49**, 261–268

Loomis, T. A. and Salafsky, B. (1963). Antidotal action of pyridinium oximes in anticholinesterase poisoning. Comparative effects of soman, sarin and neostigmine on neuromuscular function. *Toxicol. appl. Pharmac.*, **5**, 685–701

Loomis, T. A. and Johnson, D. D. (1966). Ageing and reversal of soman-induced effects on neuromuscular function with oximes in the presence of dimethyl sulfoxide. *Toxicol. appl. Pharmac.*, **8**, 533–539

Machne, X. and Unna, K. R. W. (1963). Actions at the central nervous system. *Handbuch der Experimentellen Pharmakologie. Cholinesterases and Anticholinesterase Agents.* (ed. G. B. Koelle) Springer-Verlag, Berlin, pp 679–700

Medved, L. I., Spynu, E. I. and Kagan, I. S. (1964). The method of conditioned reflexes in toxicology and its application for determining the toxicity of small quantities of pesticides. *Residue Rev.*, **6**, 42–74

Metcalf, D. R. and Holmes, J. H. (1969). EEG, psychological and neurological alterations in human with organophosphorus exposure. *Ann. N.Y. Acad. Sci.*, **160**, 357–365

Michel'son, M. I., Rozkova, E. K. and Cavateev, N. V. (1954). The antagonistic effect of cholinolytic and anticholinesterase agents on the higher nervous activities. *Bull. exp. Biol. Med.*, **2**, 7–11

Oosterban, R. A., Jansz, H. S. and Cohen, J. A. (1961). Studies with ^{18}O on the mechanism of hydrolytic enzyme reactions. *Abstracts Vth Inter. Congr. Biochem.*, Moscow, Aug 10–16, 1961, p 119.

Overstreet, D. H., Hadick, D. G. and Russell, R. W. (1972). Effects of amphetamine and pilocarpine on eating behaviour in rats with chronically low acetylcholinesterase levels. *Behav. Biol.*, 7, 217–226

Parkes, M. W. and Sacra, P. (1954). Protection against the toxicity of cholinesterase inhibitors by acetylcholine antagonists. *Br. J. Pharmac. Chemother.*, **9**, 299–305

Polak, R. L. (1963). The influence of oximes on the distribution of ^{32}p in the body of the rat after injection of ^{32}p-labelled sarin. *Acta Physiol. Pharmac. Neerl.*, **12**, 81–82

Poziomek, E. J., Hackley, B. E. and Steinberg, G. M. (1957). Pyridine aldoximes. *J. org. Chem.*, **23**, 714–717

Richardson, A. and Glow, P. H. (1967a). Discrimination behaviour in rats with reduced cholinesterase activity. *J. comp. physiol. Psychol.*, **63**, 240–246

Richardson, A. and Glow, P. H. (1967b). Post-criterion discrimination behaviour in rats with reduced cholinesterase activity. *Psychopharmacologia, Berl.*, **11**, 435–438

Rick, J. T., Morris, D. and Kerkut, G. A. (1968). Cholinesterase, cholineacteyltransferase and the production of γ-aminobutyric acid in the cerebral cortex of five behavioural strains of rats. *Life Sciences*, 7, 733–739

Riker, W. F. (1953). Excitatory and anti-curare properties of acetylcholine and related quaternary ammonium compounds at the neuromuscular junction. *Pharmac. Rev.*, **5**, 1-86

Riker, W. F. and Wescoe, W. C. (1946). The direct action of prostigmine on skeletal muscle; its relationship to the choline esters. *J. Pharmac. exp. Ther.*, **88**, 58–66

Rosenberg, P. and Coon, J. M. (1960). Nikethamide, nicotinamide and some related compounds in anticholinesterase poisoning. *J. Pharmac. exp. Ther.*, **128**, 289–298

Rosenweig, M. R., Krech, D. and Bennett, E. L. (1960). A search for relations between brain biochemistry and behaviour. *Psychol. Bull.*, **57**, 476–492

Russell, R. W. (1969). Behavioural aspects of cholinergic transmission. *Fed. Proc.*, **28**, 121–131

Russell, R. W., Vasquez, B. J., Overstreet, D. H. and Dalglish, F. W. (1971). Effects of cholinolytic agents on behaviour following development of tolerance to low cholinesterase activity. *Psychopharmacologia, Berl.*, **20**, 32–41

Russell, R. W., Warburton, D. M. and Segal, D. S. (1969). Behavioural tolerance during chronic changes in the cholinergic system. *Comm. in Behav. Bio.*, **4**, 121–128

Russell, R. W., Watson, R. H. J. and Frankenhaeuser, M. (1961). Effects of chronic reduction in brain cholinesterase activity on acquisition and extinction of a conditioned avoidance response. *Scand. J. Psychol.*, **2**, 21–29

Scaife, J. F. (1960). Protection of human red cell cholinesterase against inhibition by tabun and O,O-diethyl-S-2-diethylamino-ethyl phosphorothiolate. *Canad. J. Biochem. Physiol.*, **38**, 301–303

Schiff, M., Esmond, W. G. and Himwich, H. E. (1950). Forced circling movements (adversive syndrome). Correction with dimenhydrinate (Dramimine). *Arch. Otolaryng.*, (*Chicago*), **51**, 672–677

Sim, V. M. (1965). Chemicals used as weapons in war. *Drills Pharmacology in Medicine*. (ed. J. R. DiPalma) McGraw-Hill, New York, p 915

Smith, T. E. and Usdin, E. (1966). Formation of a nonreactivatible isopropylmethylphosphonofluoridate-inhibited acetylcholinesterase. *Biochem.*, **5**, 2914–2918

Stacek, N. K. (1959). Experimental data concerning the toxicology of methylsystox. *Pharmac. Toxicol.*, **6**, 559–570

Stacek, N. K. (1960). Data concerning the toxicological characteristic of methylsystox and the establishment of tolerance limits (Dissertation, Kiev, quoted by Medved, Spynu and Kagan, 1964)

Stavinoha, W. B., Rieger, J. A., Ryan, L. C. and Smith, P. W. (1966). Effects of chronic poisoning by an organophosphorus cholinesterase inhibitor on acetylcholine and norepinephrine content of the brain. *Adv. Chem. Ser.*, **60**, 79–88

Stoller, A., Krupinski, J., Christophers, A. J. and Blanks, G. K. (1965). Organophosphorus insecticides and major mental illness. An epidemiological investigation. *Lancet*, **1**, 1387–1388

Stratton, L. O. and Petrinovich, L. (1963). Post trial injections of an anticholinesterase drug and maze learning in two strains of rats. *Psychopharmacologia, Berl.*, **5**, 47–54

Sumerford, W. T., Hayes, W. J., Johnston, J. M., Walker, K. and Spillane, J. (1953). Cholinesterase response and symptomatology from exposure to organic phosphorus insecticides. *A.M.A. Arch. indust. Hyg.*, 7, 383–398

Thesleff, S. (1960). Supersensitivity of skeletal muscle produced by botulinum toxin. *J. Physiol., Lond.*, **151**, 598–607

Urban, E. and Sando, M. J. W. (1965). Organic phosphate poisoning. *Med. J. Austr.*, 2, 313–316

Usdin, E. (1970). *International Encyclopedia of Pharmacology and Therapeutics. Section 13. Anticholinesterase Agents.* (ed. A. G. Karczmar) Pergamon Press, Oxford, pp 47–300

Van Der Meer, C. and Wolthuis, O. L. (1965). The effect of oximes on isolated organs intoxicated with organophosphorus anticholinesterases. *Biochem. Pharmac.*, **14**, 1299–1312

Van Meter, W. G. and Karczmar, A. G. (1968). Prophylactic and antidotal treatment of sarin poisoning with drugs given singly and in combination. *Arch. int. Pharmacodyn.*, **172**, 62–72

Werner, G. and Kuperman, A. S. (1963). Actions of the neuromuscular junction. *Handbuch der Experimentellen Pharmakologie. Cholinesterases and Anticholinesterase Agents.* (ed. G. B. Koelle) Springer-Verlag, Berlin, pp 570–678

West, I. (1968). Sequelae of poisoning from phosphate ester pesticides. *Ind. Med. Surg.*, **37**, 832–836

White, R. P. (1956). Relationship between behavioural changes and brain cholinesterase activity following graded intracerebral injections of DFP. *Proc. Soc. exp. Biol. (N.Y.)*, **93**, 113–116

Wills, J. H. (1963). Pharmacological antagonists of the anticholinesterase agents. *Handbuch der Experimentellen Pharmakologie. Cholinesterases and Anticholinsterase Agents.* (ed. G. B. Koelle) Springer-Verlag, Berlin, pp 881–920

Wills, J. H. (1970). *International Encyclopedia of Pharmacology and Therapeutics. Section 13. Anticholinesterase Agents.* (ed. A. G. Karczmar) Pergamon Press, Oxford, pp 357–470

Wilson, I. B. (1951). Acetylcholinesterase. XI. Reversibility of tetraethyl pyrophosphate inhibition. *J. Biol. Chem.*, **190**, 111–117

Wilson, I. B. (1952). Acetylcholinesterases, XIII. Reactivation of alkyl phosphate-inhibited enzyme. *J. Biol. Chem.*, **199**, 113–120

Wilson, I. B. (1955). Promotion of acetylcholinesterase activity by the anionic site. *The Physical Chemistry of Enzymes,* Disc. Faraday Soc. No. 20, Aberdeen University Press, Ltd., pp 119–125

Wilson, I. B. and Ginsburg, S. (1955). A powerful reactivator of alkyl-phosphate-inhibited acetylcholinesterase. *Biochem. Biophys. Acta.*, **18**, 168–170

Wilson, I. B. and Meislich, E. K. (1953). Reactivation of acetylcholinesterase inhibited by alkyl phosphates. *J. Amer. Chem. Soc.*, **75**, 4628–4629

Zaimis, E. (1963). Actions at autonomic ganglia. *Handbuch der Experimentellen Pharmakologie. Cholinesterases and Anticholinesterase Agents.* (ed. G. B. Koelle) Springer-Verlag, Berlin, pp 530–569

5 Drug Actions at Central Muscarinic Sites

5.1 Evidence for central muscarinic receptors

Before attempting any discussion of the effects of drugs which interact with muscarinic receptors in the central nervous system it is necessary to give evidence to support the view that such receptors exist. Such evidence should show that drugs known to interact with the classical peripheral muscarinic receptors produce effects originating in the central nervous system specifically antagonised by antimuscarinic or antiacetylcholine drugs.

One set of evidence has been obtained using the technique of microiontophoretic application of drugs to single neurones in the central nervous system. This work was reviewed by Bradley in 1968 so it will not be considered here. In summary it can be stated that many neurones in a variety of areas of the central nervous system have been shown to be cholinoceptive, that is, they are excited or inhibited by acetylcholine. Of these, some satisfy the requirements for being muscarinic in that the effects are blocked by atropine. Notably, the receptor for inhibitory responses of brain stem neurones seems to be exclusively muscarinic. There is also some evidence, although it is less conclusive, that cholinoceptive cells in the hippocampus may be muscarinic.

The other sort of evidence relates to responses of the organism as a whole to muscarinic substances applied parenterally or directly into the central nervous system. The latter procedure is often necessary since most potent muscarinic agents are quaternary ammonium salts and do not penetrate into the central nervous system when given parenterally. This lack of penetration of quaternary salts is also made use of in experiments where it is required to test whether the effects of a muscarinic agonist are central or peripheral in origin. Central actions are blocked by antiacetylcholine drugs in the free base form but not when quaternised. Peripheral effects, on the other hand, are blocked by both forms of the drug.

Experiments in which drugs are given directly into the central nervous system always involve procedures which entail a certain amount of actual or potential damage to brain tissue. Consequently, interpretation of the results of such experiments purely in terms of drug actions is never entirely unequivocal. The decision here will be centred round drugs which produce their effects when given parenterally and particular emphasis will be placed on oxotremorine and related substances. Oxotremorine (see formula 18), the active metabolite of tremorine (see formula 17) is, as was discussed in Chapter 2, a muscarinic agent approximately equiactive with acetylcholine. As well as producing salivation, lachrymation, miosis, diarrhoea and bradycardia, signs typical of peripheral parasympathetic stimulation, the drug also produces, in a variety of species of animals, marked muscular tremors, muscle rigidity and akinesia. These latter effects remain unaltered if oxotremorine is given after the quaternary anticholinergic drugs methantheline or atropine methylbromide (George, Haslett and Jenden, 1962). They are most marked in rodents. Cats and monkeys display a marked sham-rage effect followed by a fear or withdrawal reaction. George, Haslett and Jenden suggested that cats and monkeys also displayed behaviour which strongly suggested hallucinatory experiences. Another central action of oxotremorine is its analgesic effect; it was stated by these authors to be 400 times as active as morphine sulphate, as determined by the guinea-pig toe pinch method.

Direct evidence for some of the actions of oxotremorine being central in origin was obtained by injecting the material to rabbits by the intraventricular route. This produced a pronounced tremor, with the dose required being one-tenth of that required to produce the same effect when given intravenously. Transection of the spinal cord in rats at T6 or L1 abolished the tremor only below the level of the section. Ablation of the midbrain tegmentum in rats prevented the tremorogenic effect of the drug but removal of structures rostral to this area did not affect the genesis of tremors. The ED50 for production of tremors in mice is about 0.05 mg/kg (Bebbington, Brimblecombe and Shakeshaft, 1966: Neumeyer, Moyer, Richman, Rosenberg and Teiger, 1967) and the ED90 about 0.07 mg/kg (Brimblecombe, Green, Stratton and Thompson, 1970).

The fact that the tremors produced by oxotremorine are blocked by tertiary base antiacetylcholine drugs but not by quaternary ammonium salts is highly suggestive that the tremors result from interaction of the drug with central muscarinic receptors. There are two additional pieces of evidence listed by Brimblecombe (1970) as follows

(1) There is a good correlation between the peripheral muscarinic activities of a series of acetylenic amines related to oxotremorine and their tremorgenic activities.

(2) The tremors are produced by tertiary amines, not by quaternary salts.

This evidence was described in some detail by Bebbington and Brimblecombe (1965) and by Bebbington, Brimblecombe and Shakeshaft (1966). Table 5.1 shows comparisons of peripheral muscarinic activity and

TABLE 5.1

A comparison of muscarinic and tremorogenic activities of some tertiary acetylenic amines (after Bebbington and Brimblecombe, 1965)

Compound	Muscarinic activity guinea-pig ileum Oxotremorine = 1	Tremorogenic activity i.v. mice Oxotremorine = 1
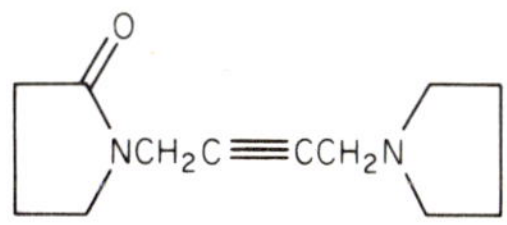	1	1
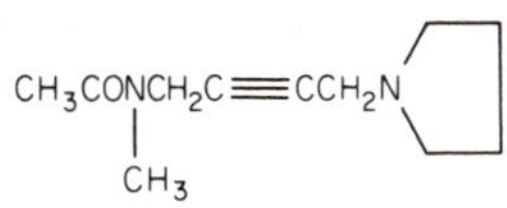	0.72	0.42
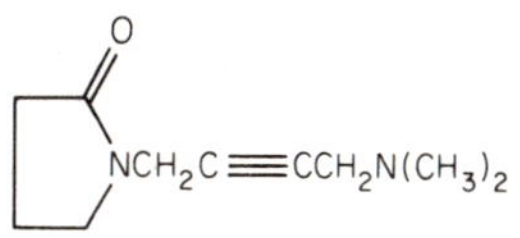	0.11	0.06
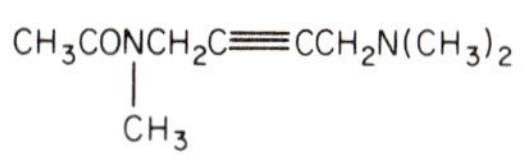	0.04	0.03
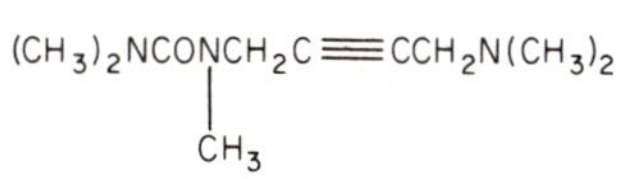	0.03	0.01
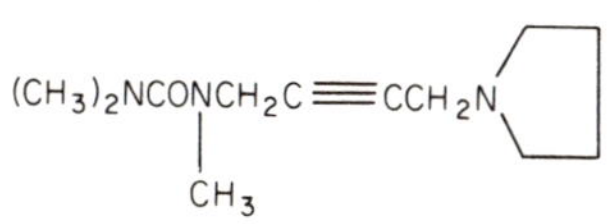	0.03	0.01
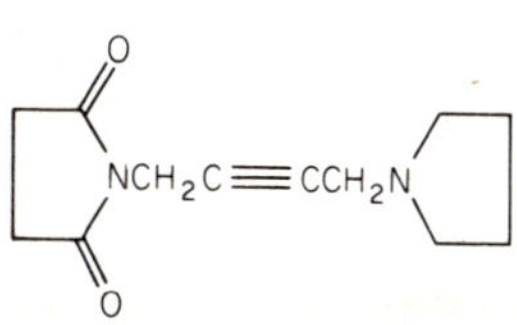	0.03	0.005

central tremorogenic activity of some tertiary acetylenic amines, including oxotremorine. It is clear that there is a general parallelism between these two activities.

In the series of investigations reported by Bebbington, Brimblecombe and their colleagues a large number of muscarinic agents of varying potencies were prepared. No quaternary ammonium salt, however potent, produced tremors.

All the above evidence, taken with that described in Chapter 2, where the structural features of oxotremorine associated with its muscarinic activity are discussed in some detail, points strongly to a direct interaction of the drug with central muscarinic receptors. There is, however, a possible alternative explanation which must be considered. Various muscarinic substances, including oxotremorine, have been shown to elevate levels of brain acetylcholine and the possibility must be considered that the drugs owe their effects to this acetylcholine which is presumably freed by their action. There are many reports of arecoline, tremorine and oxotremorine producing elevations in brain acetylcholine, an effect reversed by centrally acting antiacetylcholine drugs (Giarman and Pepeu, 1962, 1964; Holmstedt and Lundgren 1966, 1967; Cox and Potkonjak, 1969a, 1969b, 1970).

All these observations were made using the relatively non-specific bioassay for acetylcholine but they have been confirmed by Campbell, Hanin and Jenden (1970) using a specific chemical method employing gas chromatography. There has been considerable discussion concerning the mechanism whereby these elevated levels of brain acetylcholine are brought about. Holmstedt and Lundgren (1966) suggested that they could be due to increased synthesis or to increased availability resulting from mobilisation from some biologically inactive precursor. Brain homogenates incubated with oxotremorine or arecoline in the presence of physostigmine showed increased concentrations of acetylcholine. No such increase was obtained in the presence of hemicholinium which inhibits the synthesis of acetylcholine, or in the presence of atropine. This evidence would seem to favour the view that oxotremorine is stimulating the synthesis of acetylcholine but Holmstedt, Lundgren, Schuberth and Sundwall (1965) showed that oxotremorine had no effect on an *in vitro* system containing an acetone extract of cholinacetylase and Schuberth, Sollenberg, Sundwall and Sorbo (1965) showed that it had no ffect on brain acetylcoenzyme A. Since both these factors might be expected to be affected if acetylcholine synthesis was being stimulated Holmstedt and Lundgren (1966) concluded that oxotremorine was probably causing mobilisation of acetylcholine from some unidentified subcellular component.

This conclusion is not universally accepted and some confusion exists in the literature. Schuberth, Sollenberg, Sundwall and Sorbo (1966), in a later study than the one mentioned above, reported decreases in levels of brain acetylcoenzyme A after both tremorine and oxotremorine. Tremorine has reen reported to activate brain cholinacetylase both *in vivo* and *in vitro*, but a contrary finding is that of Stein and Cohen (1970) who could find no increase in brain cholinacetylase activity *in vivo* in mice after doses of oxotremorine which produced marked tremors and increases in levels of brain acetylcholine. Evidence has already been given which suggests that hemicholinium blocks the oxotremorine-induced elevation of acetylcholine. There is also a report (Slater and Rogers, 1968) that this blocking effect is prevented if the acetylcholine precursor, choline, is administered concurrently.

Thus, although there is some doubt as to the precise mechanism involved it is certain that centrally active muscarinic agents elevated levels of brain acetylcholine. The question remains as to whether this is the cause of the central effects of these drugs, or whether it is a consequence of these effects.

Earlier work by Holmstedt and Lundgren (1966, 1967) indicated that the time course of oxotremorine- or arecoline-induced tremor in animals corresponded well with the time course of the elevation in brain acetylcholine levels. Later work by Cox and Potkonjak (1969a, b, c, 1970) on the other hand, seems to indicate a lack of correlation between these time courses. These authors have also noted a lack of effect of the anticholinesterase drug dyflos on the tremors though it potentiated the rise in acetylcholine levels. They have also shown that a number of drugs can reduce oxotremorine tremor without modifying the increase in acetylcholine and vice-versa. Similarly Slater and Rogers (1968) showed that increases in brain acetylcholine levels still followed tremorine administration even when the tremor was blocked by N-(4-diethylamino-but-2-ynyl). This evidence, taken together with the fact that oxotremorine is devoid of nicotinic actions, seems to throw very considerable doubt on any theory that the drug acts indirectly by release of acetylcholine.

A recent paper by Cox and Hecker (1971) is relevant to this discussion. They studied in some detail the effects of oxotremorine on the isolated guinea-pig ileum and found that the log dose response curves for oxotremorine and acetylcholine were similar and that both drugs were competitively antagonised by atropine. The effects of acetylcholine but not of oxotremorine were potentiated by the cholinesterase inhibitor mipafox. Oxotremorine had no effect on the release of acetylcholine from the mipafox-treated ileum but decreased its release from the physostigmine-treated ileum. No evidence of an indirect component of the action of oxotremorine was found using a variety of pharmacological procedures including the use of tetrodotoxin, morphine, procaine and hemicholinium, and cooling. The authors concluded that oxotremorine acts directly on muscarinic receptors and not by releasing acetylcholine. Although it is not necessarily justifiable to extrapolate these peripheral findings to the central nervous system nevertheless it is difficult to quarrel with these authors who write "Jenden (1968) has postulated that the rise in brain acetylcholine might be due to prevention of the release of acetylcholine resulting from stimulation of a feedback mechanism by the action of oxotremorine on muscarinic receptors. There is no inconsistency between the results obtained here and the latter theory".

For the remainder of this chapter it will be assumed that agonists interact directly with receptors in the central nervous system analogous to peripheral muscarinic receptors. Even if this assumption is not correct it does not invalidate the points made, for in all probability, receptors of this type would be involved even if drug actions were mediated through acetylcholine.

It will be apparent from what has been written above that the main, and certainly the most quantitative, evidence for the presence of muscarinic receptors relates to their involvement in the production of tremors and other disorders of movement. This aspect will be dealt with in more detail in the next section of this chapter but the remainder of the chapter will be devoted to effects of agonists and antagonists on other, probably less well defined, aspects of behaviour and finally there will be a discussion of the role of central cholinergic mechanisms of the muscarinic type in behaviour.

The activities of tertiary amine muscarinic agents in inducing tremors can be used to give a measure of their central muscarinic potencies and, as a corollary, the activities of drugs in blocking the production of such tremors can be used to give a measure of their central antiacetylcholine potencies. Reference will be made to this in the section of this chapter dealing with the behavioural effects of antiacetylcholine drugs.

5.2 The involvement of central cholinergic systems in the control of movement

Antiacetylcholine drugs, in the form of belladonna alkaloids were first used in the treatment of Parkinson's disease in the middle of the last century. These naturally occurring alkaloids have now largely been superseded by synthetic antiacetylcholine drugs but such drugs still play a significant role in the treatment of the disease. This fact might reasonably be taken as an indication that the disease, or at least certain aspects of it, results from cholinergic overactivity which is antagonised by the drugs. The three main manifestations of the disease are tremor, akinesia and muscular rigidity. Antiacetylcholine drugs seem to be most effective against the latter although accounts of this vary.

The discovery that oxotremorine, which produced a syndrome very akin to Parkinsonism in various species of animals, was an active muscarinic agent and that its effects were blocked by antiacetylcholine drugs led almost inescapably to the same conclusion, that is that hyperactivity of muscarinic systems was involved in the pathogenesis of Parkinsonism.

On this basis drugs effective in blocking oxotremorine-induced tremors were regarded as candidate therapeutic agents for Parkinson's disease and, indeed, a number of useful drugs were discovered from the use of this screening method. A search was made, particularly, for drugs with preferential central actions, i.e. with higher potencies in blocking tremors than in blocking salivation.

It gradually became apparent, however, that this simple view of Parkinsonism was not justified and that the effects of oxotremorine were not precisely analogous to the manifestations of the disease. The tremor of Parkinson's disease is characteristically a rest tremor occurring alternately in opposing muscle groups and with a frequency of 4–6 Hz. The frequency of tremor produced by tremorine or oxotremorine varies considerably from species to species and, indeed, different workers have quoted differing values for the same species. Thus, the frequency of tremor in mice has been stated to be 20–24 Hz (Jenden, 1968): 12–14 Hz (Yen and Day, 1965) and 14–18 Hz (Ahmed and Taylor, 1959). In rats the frequency is generally accepted as being lower, lying between 10 and 15 Hz. As well as these reported differences in frequency it is also clear that oxotremorine-induced tremor is best described as a postural or possibly an intentional, rather than a rest, tremor and does not alternate in opposing muscle groups but occurs in phase (Decima and Rand, 1965).

The fact that oxotremorine produces a marked hypothermia as another manifestation of its central actions (Spencer, 1965) raises the possibility that the tremors may be merely shivering. This seems to be discounted by the report of Blockus and Everett (1957) that the tremor was not modified by placing the animals in a warm room.

It seems clear, then, that the tremors, and probably the other movement disorders, induced by oxotremorine, although mediated primarily through central cholinergic mechanisms, are not analogous to the disorders of Parkinson's disease. It should perhaps be added in parentheses that there is some evidence for the involvement of non-cholinergic systems in oxotremorine tremor production. In particular it has been reported by both Spencer (1966) and Morpurgo (1967) that antidepressant drugs like imipramine antagonise oxotremorine effects. Most of the other reported antagonists seem to be non-specific, producing their antagonism by counteracting the hypothermia which is produced or in the case of tremorine, affecting its biotransformation to oxotremorine.

This leaves the question of the involvement of cholinergic mechanisms in Parkinson's disease unresolved. There is no strong direct evidence for increased cholinergic activity in the disease although there have been suggestions by Jenckner and Ward (1953) for example, that lesions in the basal ganglia could lead to partial denervation of neurones in the extra-pyramidal system which might then show a denervation hypersensitivity to acetylcholine.

Such a system would clearly be amenable to antagonism by antiacetyl-choline drugs and it might be expected that the therapeutic effectiveness of such drugs would be related to their central antiacetylcholine activities. Ahmed and Marshall (1962) and Marsden (1968) concluded that this was the case, but other workers (Farquharson and Johnston, 1959) do not confirm this conclusion.

The possibility of antiacetylcholine drugs having other actions, beneficial in Parkinsonism, has been considered. Coyle and Synder (1969) found that six antiparkinsonian drugs, benztropine (see formula 110), trihexy-phenidyl (see formula 111), orphenadrine (see formula 112), diphen-hydramine (see formula 113), phenindamine (see formula 114) and

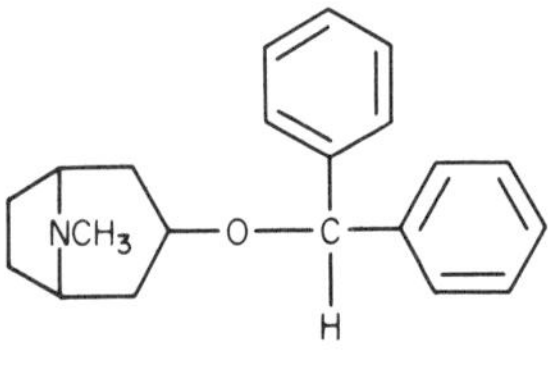

110

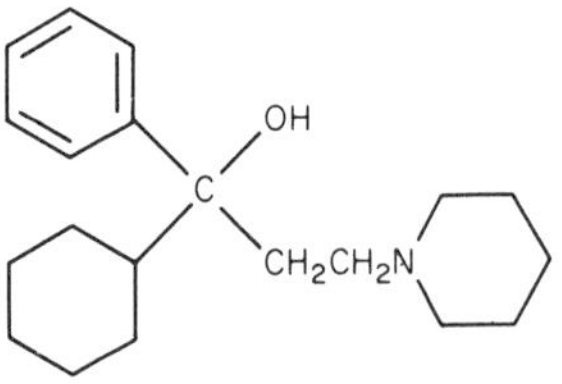

111

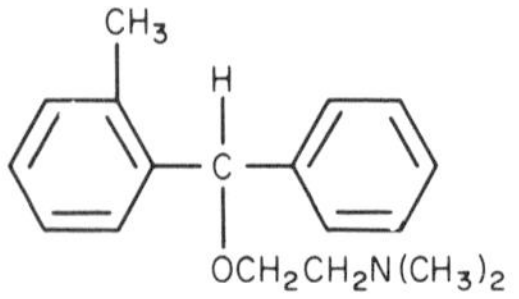

112

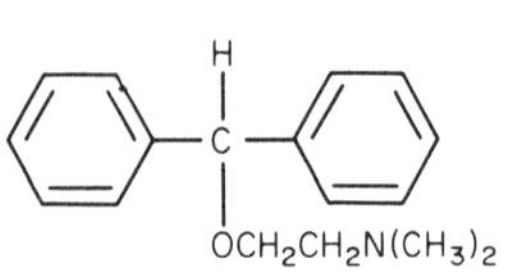

113

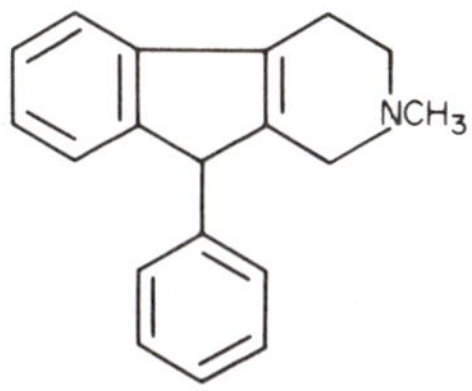

114

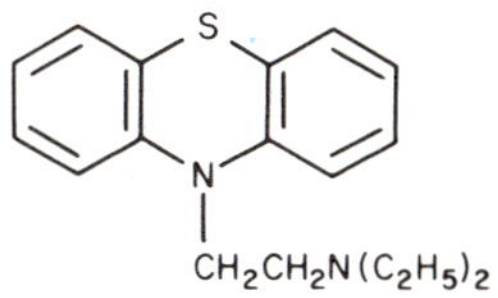

115

diethazine (see formula 115) all shared the property of inhibiting the uptake of catecholamines into synaptosomes from the striatum. Later Goldstein, Fuxe, Battista, Backstrom and Nakatani (1970) found that atropine, benztropine and other antiparkinsonian drugs all reduced the uptake of dopamine in rat striata.

The implication of these findings is that drugs, by blocking the re-uptake of amines, particularly dopamine, would potentiate the effects of these amines, an action of considerable value in conditions where their levels are reduced. It is now generally accepted that a main biochemical defect in Parkinson's disease is a decrease in levels of striatal dopamine and it is tempting to conclude that the beneficial effects of these anti-acetylcholine drugs are related to their ability to potentiate dopaminergic activity rather than to their actions on cholinergic systems. There are, however, conflicting views. Farnebo, Fuxe, Hamberger and Ljungdahl (1970), after studying the inhibitory effects of a number of antiparkinsonian drugs, including some antiacetylcholine drugs, concluded that their therapeutic efficacy could not be related to their ability to block the re-uptake of dopamine into central dopaminergic neurones.

A discussion of the role of dopamine and other amines in the pathogenesis of Parkinson's disease is not appropriate in this monograph. The subject, together with the actions of tremorogenic and tremorolytic drugs has been discussed in some detail by Brimblecombe and Pinder (1972). These authors concluded that control of motor functions involves an equilibrium between excitatory cholinergic and inhibitory aminergic systems and that disturbances of this equilibrium, produced by either drugs or disease, result in disorders of movement, including tremor. This would explain the beneficial effects of antiacetylcholine drugs and also of drugs like L-dopa which restore the dopamine deficiency. This is clearly an oversimplification of the situation but more precise analysis must await detailed information concerning the role of chemical transmitters in different brain areas.

5.3 Central actions of agonists acting at muscarinic receptors

The tremorogenic actions of oxotremorine and other muscarinic agents have been discussed in the previous section of this chapter. The present section will be concerned with other central actions of such drugs, some of which have already been mentioned in passing.

Analgesia is an effect produced by all muscarinic agents which have been examined for this particular action. Thus, tremorine (Chen, 1958), oxotremorine (Jenden, 1968) and a variety of other drugs (Herz, Holzhauser and Teschemacher, 1966) have all been reported as producing analgesia at doses below those which produce tremors. This is presumably a manifestation of a central depressant action of these drugs as is the production of hypothermia.

The ability of a variety of drugs to antagonise the hypothermia and tremors produced by tremorine was studied in some detail by Spencer (1965). The time course and severity of the two effects were very similar and they were both reduced by atropine and by other antiacetylcholine drugs like benzhexol, caramiphen and benztropine, all of proven usefulness in the treatment of Parkinson's disease. The sympathomimetic drugs amphetamine and fencamfamin antagonised both effects of tremorine with the hypothermia being more easily antagonised than the tremor. The antidepressant drug imipramine and its analogues also had definite activity against both effects but again hypothermia was more affected than tremors. Of a wide variety of drugs studied the only other two to show antagonistic activity were the neuroleptics chlorpromazine and haloperidol which were effective against tremors but not hypothermia. Their action was, however, almost certainly due to a blockade of the biotransformation of tremorine to oxotremorine and the hypothermia resulted probably from the neuroleptics themselves rather than from the tremorine.

Spencer's results seem to indicate clearly that tremors and hypothermia, while both being manifestations of central actions of tremorine, are separate effects and are not consequences of one another. Spencer suggested that the hypothermia resulted from stimulation of the rostral hypothalamus with simultaneous inhibition of the caudal hypothalamus. The rostral area, which is concerned with the promotion of heat loss, is said to be parasympathetic while the caudal area, which is concerned with heat conservation, is sympathetic (Elliot, 1963). On this basis the actions of sympathomimetic drugs in reversing the hypothermic effect of tremorine can be explained readily.

A third manifestation of a depressant action of oxotremorine is the hypoactivity it produces. Studies in the author's own laboratory in which the tremor has been measured mechanically have shown clearly that in mice given sub-tremorogenic dose levels of the drug the "tremor score" is below that of control animals that is, these animals are hypoactive.

Another muscarinic agent which has been studied for its effects on spontaneous activity is arecoline (see formula 16). Pradhan and Dutta (1970) using rats in photocell activity cages found that intraperitoneal doses of 0.25, 0.5 and 1 mg/kg caused an overall depression of motor activity. A dose of 2 mg/kg produced a significant degree of depression with approximately 50 per cent decreases in activity. If the methiodide instead of the hydrobromide was used no decrease in activity was noted indicating that the effect was central in origin. The depressant effect was antagonised by hyoscine but not by methyl hyoscine, again confirming its central origin and suggesting the involvement of muscarinic type receptors. The ganglion-blocking drug mecamylamine did not influence the effect of arecoline on spontaneous activity suggesting that nicotinic receptors were not involved.

In the same study Pradhan and Dutta investigated the effects of arecoline on operant conditioning in rats. The animals were trained to press a lever either for water or food reinforcement or for shock avoidance. Four schedules of conditioning were used, three with positive reinforcement and one with negative reinforcement. Under a fixed ratio water reinforcement schedule, doses of 1, 2 and 3 mg/kg arecoline produced a decrease in the number of reinforcements. There were two contributory factors, an increase in the average pause after reinforcement and a reduction in response rate. Under the fixed interval food reinforcement schedule, there were significant decreases in responses after 1 and 2 mg/kg arecoline. Under a schedule of differential reinforcement of low rates, doses of 0.25, 0.5 and 1 mg/kg increased the response rate and decreased the reinforcement but 2 mg/kg reduced both response rate and reinforcement. In the free-operant shock avoidance schedule, doses of 0.5, 1, 2 and 4 mg/kg of the drug resulted in a decrease in the number of avoidances and an increase in the number of schocks. The quaternary analogue of arecoline, arecoline methiodide, had insignificant and negligible effects on operant conditioning in comparison with arecoline itself.

Under the fixed ratio water and food reinforcement schedules, hyoscine did not antagonise the depression induced by arecoline, neither did mecamylamine. The lack of effects of hyoscine, as compared to its effectiveness in counteracting the decrease in spontaneous activity brought about by arecoline, seems to be due to the fact that whereas hyoscine itself tends to produce an increase in spontaneous activity, an effect opposite to arecoline, in the operant conditioning situation both drugs act in the same direction and no antagonism ensues.

The general picture to emerge from this study is one of behavioural depression mediated through central muscarinic receptors. Similar effects have been reported by a number of authors following administration of anticholinesterase drugs and some such results were referred to in the previous chapter of this monograph. However, the acetylcholine which accumulates as a result of the action of anticholinesterase drugs would have both muscarinic and nicotinic activity and further analysis of the relative effects on behaviour of these two activities was made by Olds and Domino (1969) using the technique of self-stimulation in rats. These rats were trained in Skinner boxes to press a lever to deliver brief electric shocks to the hypothalamus via chronically implanted indwelling electrodes. Three drugs were studied, physostigmine, arecoline and nicotine. Both physostigmine (subcutaneous doses of 50 mg/kg and above) and arecoline (100 mg/kg and above) depressed self-stimulation, an effect antagonised by hyoscine but not methylhyoscine and prolonged to some extent by mecamylamine but unaffected by trimethidinium. The time to onset of the effects of physostigmine was much longer than with arecoline presumably because with the former, time was required for sufficient inhibition of cholinesterase to occur to allow consequent accumulation of adequate quantities of acetylcholine. Recovery from the effects of physostigmine was much more abrupt than from arecoline. Nicotine also depressed self-stimulation but its effects were transient and variable and the depression was sometimes followed by facilitation. The depressant effects were blocked by both hyoscine and mecamylamine but not by methylhyoscine or trimethidinium.

These findings, then, are also consistent with the hypothesis that there is a central muscarinic inhibitory system. The effects of nicotine, which are less straight-forward, are discussed in more detail in the next chapter, but one finding of interest which requires comment is the prolongation of the effects of physostigmine and arecoline by the nicotinic antagonist mecamylamine. According to Olds and Domino this phenomena would be expected on the basis of a hypothesis proposed by Pfeiffer, Beck and Goldstein (1967). These workers suggested that a balance exists between muscarinic and nicotinic functional groups of the cholinergic receptor with a block of one part of the receptor enhancing the interaction of an agonist with the other part. The suggestion was that the balance lay between the methyl-oxygen-oxygen (M-O-O) and nicotinic onium head (N^+). So mecamylamine, by blocking the N^+ receptor site, would enhance the actions of arecoline and physostigmine on the M-O-O site thus augmenting the behavioural depression.

Whether this explanation is correct is open to argument. There may be more reason to suppose that completely separate muscarinic and nicotinic receptors are involved but, in any case, the concept of muscarinic and nicotinic systems in balance seems a reasonable one to put forward at the present time and should form the basis of further experimental work.

5.4 Central actions of antiacetylcholine drugs

5.4.1 *Mode of action*

A good deal of the material in the remainder of this chapter will be concerned with an investigation of the hypothesis that behavioural and electrophysiological effects of antiacetylcholine drugs are brought about by these drugs acting centrally as competitive antagonists of acetylcholine at muscarinic receptors. The evidence for and against this point of view will be discussed later, but it is necessary at the onset to note that there have been many reports of antiacetylcholine drugs causing decreases in the total acetylcholine content of the brain. This was shown, for example, by Giarman and Pepeu (1962) and there have been suggestions (Giarman and Pepeu, 1962; Giarman and Freedman, 1965) that the psychotomimetic activities of such drugs might be directly related to these alterations in brain acetylcholine levels. It is difficult to obtain direct evidence to test this suggestion but on present evidence its seems likely that these decreased levels of brain acetylcholine are the consequence of cholinergic blockade, that is, that they result from a feedback mechanism analogous to, or perhaps identical with, that proposed by Jenden (1968) to explain the elevations in brain acetylcholine levels which follow oxotremorine administration. Even if this is the case it does not detract from the view that the decreased levels of acetylcholine might produce the behavioural effects of these drugs. Evidence presented in the remainder of this chapter, however, would seem to suggest that the drugs act by direct, rather than by such indirect, effects.

Abood (1968, 1970) made a number of suggestions concerning the mode of action of antiacetylcholine drugs. He noted, in the first instance, that various esters of glycollic acid were capable of inhibiting the spontaneous twitching which occurs when isolated frog sartorius muscles are immersed in calcium-free Ringer's solution. Later he and his colleagues found that partial depolarisation of the muscle which occurred upon immersion in the calcium-free medium could be reversed by the addition of low concentrations of certain glycollates to the solution. This implied that the glycollate was acting in a manner similar to calcium ions in restoring the resting potential of the muscle. Another similarity between the action of the drugs and of calcium ions was observed on excited single muscle fibres. Both an increase in the concentration of calcium in the Ringer's solution and the addition of the glycollate produced a marked prolongation of the negative after-potential, an effect presumably due to a decreased permeability of the membrane to ions like Na^{+} and K^{+}. From this it was concluded that the glycollates were similar to Ca^{++} ions in their effects on the excitable membrane. It is possible that these membrane effects contribute to the central actions of the glycollates, including their effects on behaviour. Certainly, if as Abood suggests, these drugs occupy Ca^{++} binding sites in an irreversible fashion this might be expected to produce a nerve block and the fact that the drugs have local anaesthetic activity is not incompatible with such a suggestion.

Another suggestion made by Abood was that a carbonium ion mechanism might be involved in the mode of action of the psychotomimetic glycollates; for example

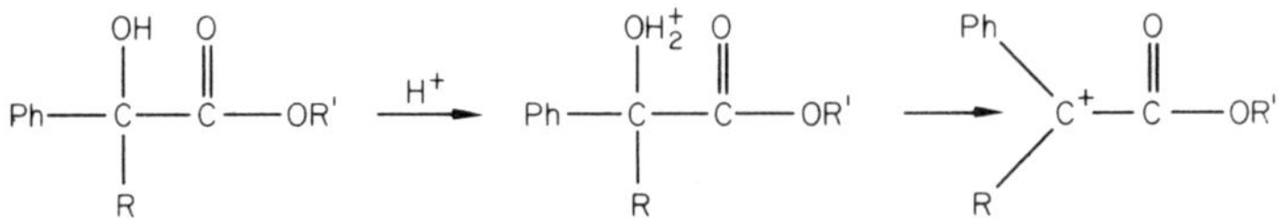

The formation of a carbonium ion would, necessarily, destroy the asymmetry of the molecule and it would not, therefore, be expected that stereochemical factors would influence activity. However, it has been shown by Brimblecombe, Green, Inch and Thompson (1971) that considerable differences in both peripheral and central antiacetylcholine potency do exist between the enantiomers of compounds of this type. Some results to illustrate this point are given in Table 5.2

Reference to the results in this table also seems to invalidate another suggestion made by Abood, that the duration of action of antiacetylcholine drugs is related to their ease of acidic hydrolysis. It can be seen that pairs of enantiomers which are hydrolysed at the same rate have quite different durations of action.

From the results reported by Brimblecombe *et al.* (1971) and from many other sources it is clear that quaternary compounds are usually more potent as antiacetylcholine drugs than the corresponding tertiary compounds. This implies that the contribution of the lone pair of electrons on the nitrogen is of relatively little importance in the development of antiacetylcholine activity and seems to be at variance with Abood's suggestion that the antiacetylcholine activity of glycollic acid esters of tertiary aminoalcohols is very dependent on the availability of this lone pair of electrons, especially in cyclic aminoalcohols.

If the possibility, discussed in Chapter 3 of this monograph, that these drugs might interact with receptors distinct from the receptor for agonists is added to the above considerations then it is apparent that much remains to be elucidated concerning their central, and to some extent their peripheral modes of action.

TABLE 5.2

Peripheral and central potencies of the racemates and enantiomers of two anticholinergic drugs (from Brimblecome *et al.*, 1971)

Drug		Antagonism of carbachol-induced contractions of the isolated guinea-pig ileum $\log K_B \pm$ S.E.	Production of mydriasis in mice: Potency relative to atropine with 95% limits	Production of mydriasis in mice: Time to peak response (min)	Production of mydriasis in mice: Duration of effect (min $\pm$ S.E.)	Antagonism of oxotremorine effects in mice ED_{50} (μmol/kg): Salivation	Antagonism of oxotremorine effects in mice ED_{50} (μmol/kg): Tremors
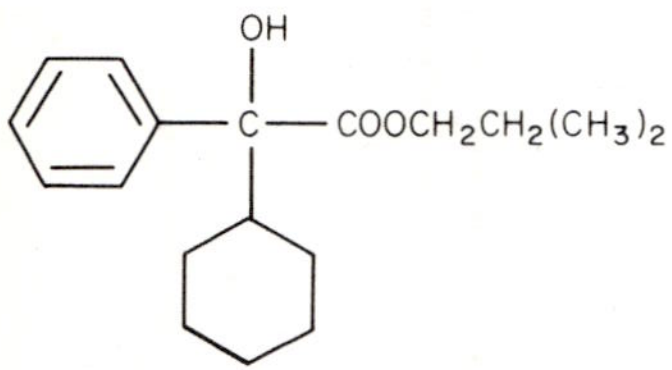	R enantiomer	9.06 $\pm$0.01	0.76(0.56–0.96)	10	73 $\pm$3	0.76(0.44–1.29)	5.07(3.87–6.9)
	Racemate	8.73 $\pm$0.02	0.34(0.28–0.41)	10	58 $\pm$5	1.40(1.11–3.33)	22.7(9.39–41.4)
	S enantiomer	7.07 $\pm$0.01	0.006(0.004–0.008)	< 5	65 $\pm$6	> 100	> 100
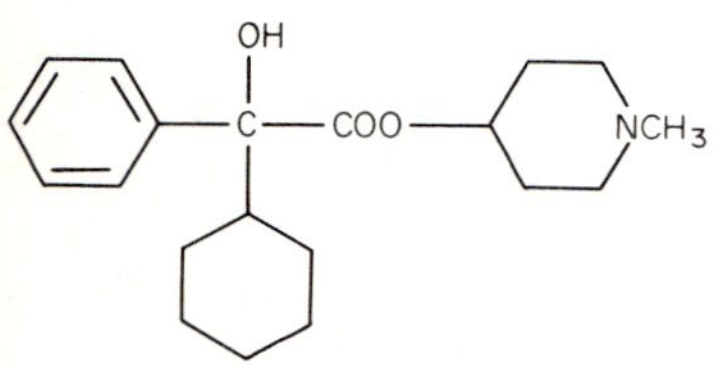	R enantiomer	10.92 $\pm$0.10	2.36(2.04 – 2.72)	80	901 $\pm$85	0.18(0.10–0.31)	0.56(0.30–1.3)
	Racemate	10.53 $\pm$0.07	1.17(1.02–1.36)	90	552 $\pm$45	0.31(0.18–0.54)	3.29(1.35–8.44)
	S enantiomer	8.48 $\pm$0.04	0.12(0.10 $\pm$0.14)	5	32 $\pm$4	7.75(3.8–16.2)	9.42(5.54–16.6)

5.4.2 Effects on man

The fact that antiacetylcholine drugs can produce profound behavioural disturbances in man has been recognised ever since it was appreciated that these drugs were responsible for the effects of belladonna poisoning. Since that time there have been appreciable numbers of recorded incidents of toxic psychoses resulting from homatropine eye drops, belladonna plasters, sleeping preparations containing hyoscine and so on. No attempt will be made here to review in any detail the literature relating to this topic. A succint review was made by Longo (1966) and he concluded that the following symptoms of atropine poisoning had been reported consistently

(1) Diminution of the power of concentration and memory
(2) Drowsiness, withdrawal and tendency to sleep
(3) An excitatory effect
(4) Loss of coordination of voluntary movements (ataxia) and inability to carry out complex actions (asynergia)
(5) Hallucinations.

The latter effect seems only to occur after large doses and so it is probably not appropriate to describe atropine as an hallucinogenic drug. Hyoscine has been quite widely used as a sedative drug but it seems clear that once a critical dose is exceeded then its effects are barely distinguishable from those of atropine.

Apart from these naturally occurring alkaloids there are also a number of synthetic antiacetylcholine drugs which produce marked psychotomimetic effects. Notable among these are the glycollates (see formula 116) first reported by Abood, Ostfeld and Biel in 1958

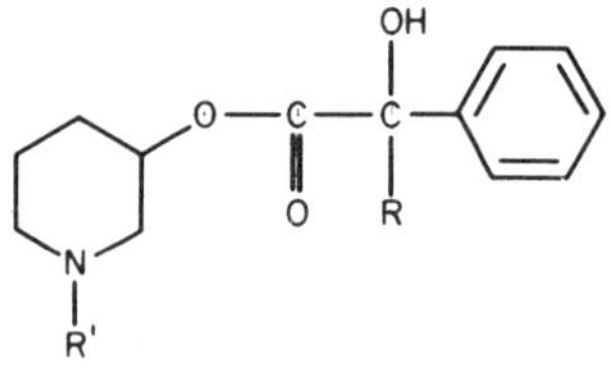

116

The effects of these drugs in man are complex as they are dose-and agent-dependent and also vary with the psychological make-up of the subject. The detailed effects of these drugs are listed by Abood (1968). The drug about which most information is available is Ditran (see formula 98) and Abood lists the psychic-behavioural effects of this drug given, by a variety of investigators, in doses of between 1 and 25 mg by the oral, intra-muscular or intravenous routes. Characteristically in the period 5 to

15 min after intramuscular injection autonomic effects are seen with mydriasis, dryness of the mouth, tachycardia, hyper-reflexia, muscle weakness and dizziness. These effects are accompanied by what Abood describes as emotional-affective symptoms, that is, apprehension, feeling of intoxication, giddiness, euphoria, fear and depersonalisation. There then follow in turn signs of intellectual impairment and perceptual changes characterised respectively by reduction in the speed and accuracy of problem solving and impairment of concentration and by distorted images, illusions, hallucinations, body image distortion etc. During the period 30 to 60 minutes after injection of the drug a stuporous state may develop with incoherence, somnolence and disinterest. Delirious states have also been reported as occurring between 45 and 120 minutes after drug administration with confusion, restlessness and agitation, marked motor and mental incoordination and aimless motor and mental activity. During the same period effects of an apraxic type may also appear with "impaired motor, amnestic, ideational, and ideomotor abilities". Between 1 and 2 hours after the drug there is evidence of occurrence of aphasia and agnosia with senseless, rambling talking, loss of ability to concentrate and misinterpretation of questions. Emotional signs have also been noted during this period with sudden outbursts of anger or other emotions, paranoia, panic and fear. It is generally agreed that these drugs produce amnesia. Abood quotes the time of onset for partial or complete amnesia as being 1 to 8 hours after injection of the drug.

This description, paraphrased from Abood's detailed summary, is sufficient to indicate the profound behavioural changes produced by drugs of this type. Abood also suggests that there seems to be a close similarity between the spectrum of symptoms seen in schizophrenia and those elicited by the glycollate esters. The parallelism is not, however, complete. The panic, degree of bewilderment and confusion and intense sensorial clouding often seen after the drugs are not normally associated with schizophrenia. However, there have been several claims that Ditran and other glycollates produce a schizophrenia-like syndrome (Meduna and Abood, 1959; Ostfeld and Biel, 1959: Gershon and Olarin, 1960). In view of this there is interest in a study by Neubauer, Gershon and Sundland (1966) which was concerned with the effects of Ditran on a mixed psychiatric population. The population was divided into five groups on the basis of diagnosis and all the subjects, 33 in all, were given 0.05 mg/kg Ditran intravenously. The response to the drug was evaluated using a Ditran rating scale which contained 80 factors which had shown changes after Ditran in previous investigations. The subjects also copied Bender-Gestalt (B-G) figures.

The general effect of Ditran was to produce a central nervous system depression, neurological regression and clouding of consciousness being the most marked effects. The most apparent differential effect was the relative lack of response of the groups diagnosed as being actively schizophrenic. This finding applied to both behavioural and autonomic effects of the drug and seems to be in agreement with results obtained by other workers using different psychotomimetic agents. It was pointed out, for example, by Huxley, Mayr, Osmond and Hoffer (1964) that there was the possibility of schizophrenia being a genetic morphism and these authors noted the rather general resistance shown by overt schizophrenics to a variety of drugs.

Another finding in the study by Neubauer, Gershon and Sundland was that patients in their chronic inactive group, who all showed diminution of drive, flattening of effect and personality deficit, tended to be behaviourally and neurophysiologically activated following Ditran administration. The authors suggested that the drug may have some therapeutic value in such patients. More recently Itil, Keskiner and Holden (1969) investigated the use of Ditran and also LSD-25 in the treatment of schizophrenics who over a long period had proved to be resistant to treatment with conventional psychotherapeutic agents. The choice of drug to be given to the patients was made on the basis of their EEG patterns. Nine patients with predominantly hypersynchronised alpha EEGs were given LSD-25 (4–15 injections of 5 to 84 μg) and eight patients with predominantly rhythmical patterns and/or slow activity were given Ditran (4–15 injections of 1 to 21 mg). Six patients in each group completed the study period during which time treatment with LSD-25 on Ditran was continued until the EEG pattern was modified or the behaviour pattern deteriorated. At this stage conventional psychotherapeutic drug treatment was reinstated.

With reference to the results in patients given Ditran, it transpired that four out of six subjects in this group showed deterioration in behaviour following Ditran and that two of these four together with the two not markedly affected by Ditran were sufficiently improved at the end of their post-Ditran period of treatment with a conventional psychotherapeutic drug to be discharged from hospital. The remaining two patients also showed improvements. It was noted that Ditran particularly activated the symptoms relating to mood, effect and consciousness. The drug also desynchronised the stable hypersynchronous EEG patterns in these patients and it seemed that those who exhibited the most marked EEG changes were those who responded best to subsequent conventional drug therapy. With LSD-50 three out of six patients had recovered sufficiently to be discharged from hospital at the end of the procedure. The behavioural effects of LSD-25, in contrast to those of

Ditran, were mainly manifested in exaggerated perceptual disturbances, disorders of thought content and thought process.

There seems to have been little follow-up of such use of Ditran in this so-called symptom provocation approach but there have been a number of studies of Ditran in the treatment of depression. This followed an early observation by Abood and Meduna (1958) that some subjects showed improvements in mood and motivation after the drug. Subsequently, studies by, for example, Finkelstein (1961) and English (1962) indicated that the drug had some value in the treatment of certain forms of depression and, although it has not been widely used for this purpose, it does raise the question as to whether the anticholinergic actions of the drug are responsible for its antidepressant activity. This is particularly relevant in view of the finding by Rathbun and Slater (1963) that the tricyclic antidepressant drugs amitriptyline and nortriptyline have marked antiacetylcholine actions in both the peripheral and the central nervous system and the suggestion by Cairncross, Gershon and Gust (1963) that imipramine might, by virtue of its antiacetylcholine actions, reduce excessive cholinergic activity which seems to be related to some depressive states.

This led to Brimblecombe and Green (1967a) to study some peripheral and central antiacetylcholine actions of amitriptyline, imipramine and their demethylated derivatives and, in particular, to ascertain whether their effects differed quantitatively or qualitatively from those of other antiacetylcholine drugs. All the drugs inhibited acetylcholine-induced contractions of the isolated guinea pig ileum, their potencies in this respect being one to two orders of magnitude lower than atropine. Amitriptyline also antagonised oxotremorine effects in mice. Probably the most interesting results related to the activities of the drugs in elevating the electrocortical arousal threshold in cat *encéphale isolé* preparations and in cats with chronically implanted cortical electrodes. All the drugs produced a dissociation between electrocortical and behavioural arousal thresholds, an effect discussed in more detail in the next section of this chapter. The pattern of the dissociation resembled that produced by hyoscine, rather than by atropine or the glycollates, in that there was a marked sedative component in the action of the drugs. The demethylated compounds desmethylimipramine and nortriptyline, were less potent than the parent compounds both in terms of their peripheral antiacetylcholine potencies and their potencies in elevating the electrocortical arousal threshold: Clinically there is no evidence to suggest that the demethylated compounds are less active as antidepressants than the parent compounds and it was concluded that clinical antidepressant activity is not primarily associated with the antiacetylcholine actions of these drugs although these actions might account in part for the qualitative differences which exist between these tricyclic antidepressants and other antidepressants of the monoamine oxidase type for example.

The general picture to be gained from the fore-going accounts is that antiacetylcholine drugs produce a characteristic behavioural syndrome which differs in a number of respects from that produced, for example, by LSD-25. This matter was studied in detail by Isbell, Rosenberg, Miner and Logan (1964) who carried out human experiments with hyoscine, N-ethyl-3-piperidyl benzilate and LSD-25 to obtain information on the development of tolerance and cross tolerance to and between these drugs. There was clear development of direct tolerance to all three drugs on most of the parameters which were measured. The results on cross tolerance were, however, not so clear cut. As might have been anticipated there was a considerable degree of cross tolerance between the two antiacetylcholine drugs but this did not include all the effects produced. Again, as was expected there was little or no evidence of cross tolerance between LSD-25 and the antiacetylcholine drugs. So although the results of this study were not unequivocal they did tend to confirm the view that LSD-25 and antiacetylcholine drugs act by different mechanisms to produce their psychotomimetic effects.

From the above account it is apparent that antiacetylcholine drugs produce, in man, behavioural changes perhaps best described as a toxic psychosis. The syndrome shows some similarity to schizophrenia though there is not a precise analogy between the two. In addition to this effect there is some suggestion that some drugs of this type are effective in reversing some types of depression although even this is done at the expense of producing a psychotic type of reaction. Whether or not these behavioural effects can be related directly to the antiacetylcholine actions of the drugs is still not resolved and probably is not capable of resolution until much more quantitative information is available concerning the activity of these drugs in man. With such information it should be possible to attempt to demonstrate correlations between antiacetylcholine potency and psychotomimetic activity. In the meantime the accumulation of information from animal studies should go some way towards answering the questions. The main difficulty in this respect lies in relating objective behavioural or electrophysiological changes in laboratory animals to the largely subjective behavioural effects in man.

5.4.3 Effects on electrical activity of the brain

The effects of antiacetylcholine drugs on brain electrical activity have been widely investigated but the results of all these investigations are not in agreement. A few authors have described only slight and inconsistent changes. For example Wilson (1961) could find little effect on the EEG in man following toxic doses of atropine, a result consistent with that of Mellerio (1964). More recently Wilson and Hughes (1964) have stated that atropine produces slowing of the EEG rhythm and that Ditran, in hallucinogenic doses, often produces disappearance of alpha activity. Even these latter effects are, however, relatively slight, and inconsistent.

Another series of papers reports increases in high frequency EEG activity with desynchronisation but many of these relate to the actions of the drugs on the abnormal EEG. For example Ulett and Johnson (1957) showed that delta activity which was observed during the administration of electroconvulsive therapy was blocked by atropine and hyoscine and Fink (1960) found the same effect with a wider variety of antiacetylcholine drugs. Lechner (1956) reported a tendency for hyoscine to normalise the EEG in patients with concussion.

The majority of reports (see for example, Gibbs, Gibbs and Lennox (1937); Grob, Harvey, Langworthy and Licenthal (1947)), however, seem to describe increases in low voltage slow-wave activity and a reduction in amplitude of the 8 to 10 Hz alpha waves.

A most interesting finding was reported by Wikler in 1952. He noted that dogs given atropine had an EEG pattern with high voltage slow waves and spindle activity, normally characteristic of sleep, irrespective of whether the animals were quiet and apparently dozing or awake and barking or struggling. Wikler termed this phenomena an EEG—behavioural dissociation referring to the failure of the EEG pattern to shift in the expected direction. So Wikler concluded that, although the symptoms subserving synchronisation and desynchronisation of EEG were normally interlocked with those systems subserving sleep and wakefulness, they were distinct, and that sleep or wakefulness were not consequences of changes in electrocortical activity. In a later paper Wikler (1954) extended these conclusions to suggest that the systems subserving synchronisation and desynchronisation of the EEG were also independent of the systems involved in ideation, mood level of awareness, motility and sensation. Wikler gave a succinct summary of his views in 1968. Since Wikler's original paper there have been numerous reports of this kind of dissociation in other species and with other drugs. These reports are listed and discussed by, among others, Bradley (1968), Ostfeld, Machne and Unna (1960), Itil and Fink (1968) and Longo (1966). It should be emphasised, however, as it was by Wikler (1968) that the dissociation between EEG and behaviour is not necessarily sustained throughout the period of action of the drug. Indeed, in his original paper Wikler reported alternating periods of association and dissociation.

A surprising amount of controversy has been generated over the years concerning this matter of dissociation. This has arisen mainly because a number of workers have attempted to use the term for phenomena distinct from the one described by Wikler. Unquestionably if certain aspects of behaviour are studied this dissociation does not exist but often

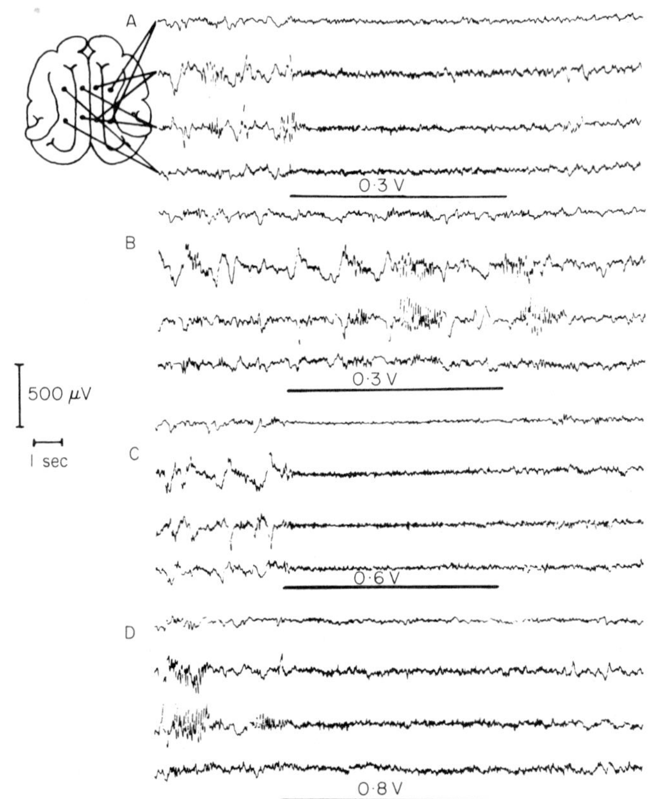

FIGURE 5.1 *Effect of atropine sulphate on EEG arousal produced by stimulation of the brain stem reticular formation in the cat* encéphale isolé *preparation*
A Control arousal (0.3v stimulus)
B After 1.2 μg/kg atropine sulphate i.v. (0.3v stimulus)
C After 1.2 μg/kg atropine sulphate i.v. (0.6v stimulus)
D After 2.4 μg/kg atropine sulphate i.v. (0.8v stimulus)

it cannot exist because there is no EEG pattern consistently associated with the behaviour being studied.

The type of dissociation described by Wikler is quantifiable by a method first described by Bradley and Key (1958). The technique involves the use of the cat *encéphale isolé* preparation. Behavioural and electrocortical arousal can be produced in this preparation, which is normally sleepy, by electrical stimulation of the brain stem reticular formation (see figure

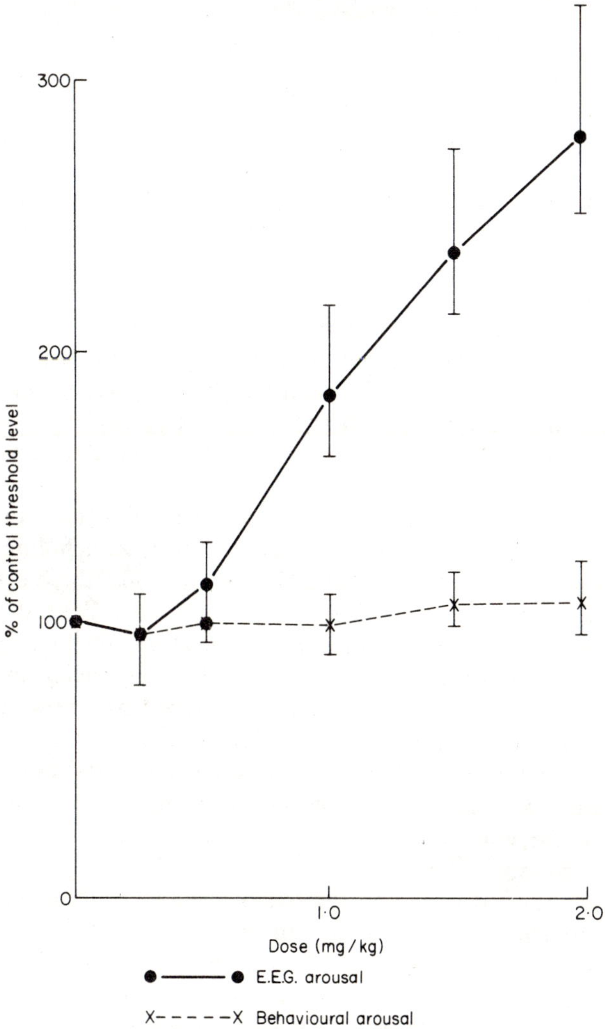

FIGURE 5.2 *Effect of atropine sulphate given i.v. to the cat* encéphale isolé *preparation on thresholds for EEG and behavioural arousal produced by stimulation of the brain stem reticular formation.*

5.1). Atropine and related drugs produce an elevation of the threshold for electrocortical arousal whereas the threshold for behavioural arousal remains unchanged.

This effect can be displayed graphically (see figure 5.2) and it is clear that until a plateau level is reached there is a progressive increase in the elevation of the electrocortical arousal threshold with dose.

The question arises as to whether the antiacetylcholine drugs produce this effect by virtue of their action as competitive antagonists of acetylcholine at central cholinoceptive sites. The qualitative evidence for this is good in that all antiacetylcholine drugs produce the effect and it can be reversed by the administration of anticholinesterase agents (see figure 5.1).

Rinaldi and Himwich (1955 a, b) suggested that the brain stem reticular formation together with the thalamic diffuse projection systems might act together as a system which they called the mesodiencephalic activating system to bring about the EEG arousal reaction. They furthermore suggested, on the basis of the kind of evidence given above, that this reaction is cholinergic in nature. Studies by Smirnov and Ilyutchenok (1962) and Cucilic, Bost and Humwich (1968) in which antiacetylcholine drugs were applied typically to the cortex suggested that cholinergic structures close to the cortical surface constituted the final corticopetal link in EEG alerting.

Brimblecombe and Green (1967b, 1969) were able to demonstrate a quantitative relationship between the central antiacetylcholine potencies of a series of drugs and their potencies in elevating the electrocortical arousal threshold in cat *encéphale isolé* preparations. Central antiacetylcholine potency was expressed in terms of the ability of the drug to block oxotremorine-induced tremors in mice, while potencies in elevating electrocortical arousal thresholds were expressed as doses required to elevate this threshold by 100 per cent. Those doses were read off from graphs of the sort shown in figure 5.2. There was a statistically significant correlation between these values, as shown in figure 5.3, thus providing quantitative evidence for a cholinergic link in the mesodiencephalic activating system.

As was mentioned earlier in this chapter there is abundant evidence that antiacetylcholine drugs reduce the brain acetylcholine content and there have been suggestions that these reduced acetylcholine levels are responsible for at least some of the behavioural effects of these drugs. Other authors have suggested that the EEG changes seen after antiacetylcholine drug administration can be explained similarly. Thus Matin (1969) studied the effects of atropine, hyoscine and total alkaloids of *Datura alba* on the EEG and total acetylcholine content of the whole brain of rats. The drugs were all given in doses of 25 mg/kg by the intraperitoneal route and EEG recordings taken from frontal and parietal lobes. In control animals the EEG frequency ranged from 9 to 12.5 Hz and the amplitude from 20–25 μV. The total brain acetylcholine content was 2.68 μg/g brain tissue. After atropine the frequency range dropped to 7.5 to 9.5 Hz and the voltage rose to 35 to 45 μV. The acetylcholine content was 1.75 μg/g. The corresponding value in rats given hyoscine were 6.5 to 8 Hz, 35–50 μV and 1.58 μg/g acetylcholine. After total *Datura* alkaloids the figures were 5 to 6 Hz, 40–55 μV and 1.14 μg/g. Clearly there is some parallelism between increase in voltage and reduction in frequency of the EEG and reduction in brain acetylcholine content. Whether this is a causal relationship or merely two separate consequences of some primary effect it is impossible to say.

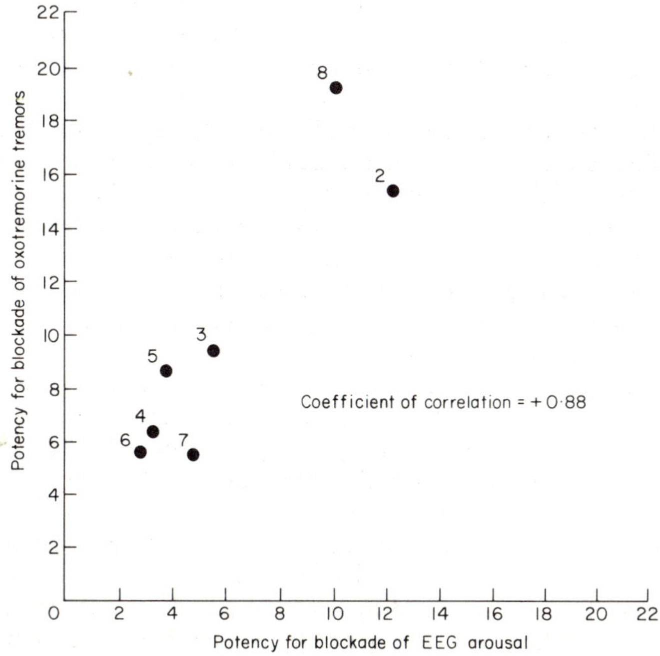

FIGURE 5.3 Relationship between the potencies of drugs in blocking oxotremorine-induced tremors and in elevating the EEG arousal threshold. All potencies expressed relative to atropine. Compound: 2 = hyoscine, 3 = N-methyl-3-piperidyl-phenylcyclopentylglycollate, 6 = N-ethyl-3-piperidyl-phenylcyclopentylglycollate, 7 = N-ethyl-2-pyrrolidylmethyl-phenylcyclopentylglycollate, 8 = Ditran (70% of 7 + 30% of 6).

One possible criticism of most of the studies in which dissociation between behaviour and EEG has been demonstrated is that the assessment of the state of the EEG has been made purely by visual examination of the records.

It has been suggested (Fink, 1968) that more extensive use may be made of computer methods for EEG analysis or that evoked potentials should be studied. A study of the latter kind has recently been reported by

Green and Aldous (1971). They used three antiacetylcholine drugs, atropine, hyoscine and N-methyl-3-piperidyl benzilate (see formula 116, R = Ph, R′ = CH_3) and investigated their effects on EEG, behaviour and evoked auditory potentials in cats. The EEG was recorded from auditory and association areas of the cortex and evoked potentials from the auditory cortex and lateral lemniscus.

In an initial series of experiments intramuscular doses of 2.4 mg/kg atropine sulphate, 0.24 mg/kg hyoscine hydrobromide and 0.5 mg/kg N-methyl-3-piperidyl benzilate hydrochloride were used. All produced synchronisation of the EEG but no gross changes in behaviour. The main finding with regard to evoked potentials recorded from the lateral lemniscus was that the fluctuations in their amplitude when recorded under distractive conditions were smaller than those in control animals under similar circumstances. The implication of this finding seems to be that the antiacetylcholine drugs, in contrast to LSD-25, decrease the significance of stimuli and hence decrease the degree of distractibility of the animals. Green and Aldous suggest that this may be related to the "introvertive" state produced in man by antiacetycholine drugs in comparison to the "extrovertive" state produced by LSD-25.

The typical surface potential evoked on the auditory cortex of these cats in response to a tonal pip comprised two short latency primary potentials (P_1 and P_2) followed by a negative potential (N_1) (Teas and Kiang, 1964) and then by a large latency secondary positive potential (+S). The antiacetylcholine drugs produced a fusion of P_1 and P_2 potentials so that subsequent amplitude analysis of these individual waves was impracticable and the maximum height of the primary complex (+P) was analysed together with the amplitude and peak latency of the +S potential. The effect of the drugs in fusing the P_1 and P_2 components is of some interest however. It is generally agreed, following the work of Teas and Kiang that the P_1 wave reflects the arrival of the nerve impulse through the primary sensory afferent pathway and the excitation of neurones in the deeper layers of the cortex. The P_2 wave is less well understood but Anochin (1961) suggested that it is produced by several corticopetal pathways. If this is so the fusion of P_1 and P_2 into a single wave must be due to an action of the antiacetylcholine drugs at the cortical level.

The effect of the antiacetylcholine drugs in animals kept under either sound-proof or distractive conditions was to increase the mean amplitude of the +P complex, to decrease the amplitude of +S and, most markedly, to increase the peak latency of the +S potential. It has been shown by Key (1965) that, in cats, the amplitude of the +S component of the evoked potential in the auditory cortex is closely related to the significance of the stimulus at least in terms of its effect in producing arousal or gaining attention. The potential could be the result of the arrival of impulses in deeper layers of the cortex via collateral routes including the brain stem reticular formation and non-specific thalamic structures. Szerb showed in 1965 that topically applied atropine reduces the amplitude of the long latency secondary potential in the somatosensory cortex and this finding makes it probable that the antiacetylcholine drugs in the study by Green and Aldous were acting at the cortical level to increase the peak latency of the +S potential.

It has been noted that, despite the described marked effects of the antiacetylcholine drugs on both the EEG and evoked auditory potentials, there were no apparent effects on the overt behaviour of the animals. In another series of experiments, however, Green and Aldous investigated the effects of the same drugs on a conditioned avoidance response in cats. The animals were conditioned to respond to a buzzer by crossing a barrier in a sound-proof chamber. Failure to cross was punished by an electric shock delivered via the grid floor of the chamber. Evoked auditory potentials and EEGs were recorded from these animals. Since the initial findings suggested that the antiacetylcholine drugs were reducing the significance of stimuli it might have been expected that this type of conditioned behaviour would be disrupted and, in fact, all three drugs at the doses mentioned above produced a block of the conditioned avoidance response. The EEGs in these animals were always synchronised in comparison with control animals where, during the test procedure, there was always desynchronisation of the EEG corresponding with the degree of alertness or attentiveness of the animals. This fact complicates any interpretation of the evoked potentials recorded from these animals since it is well known that, at any rate in the normal animal, the waveform of evoked potentials varies with the type of EEG background upon which they are superimposed. However, the effects of the antiacetylcholine drugs were to increase the amplitude of the +P potential and to decrease that of the +S potential. There was also a tendency for the amplitude of the N_1 potential to be reduced but this effect was not consistent and in any case the neurophysiological mechanisms involved in the generation of this potential are not really understood (Ochs, 1968). There was a tendency for the amplitude of evoked potentials in the lateral lemniscus to be reduced but this effect was not statistically significant. These findings, then, are not at variance with the view that the effects of the antiacetylcholine drugs in decreasing the amplitude and increasing the peak latency of the +S potential are related to their effects in reducing the significance of stimuli and hence in disrupting conditioned behaviour. Green and Aldous maintain that this disruption of the conditioned avoidance response at a time when the EEG shows a preponderance of high voltage slow waves is indicative of an "association" rather than a "dissociation" between EEG and behaviour. This, however, seems to be taking the term "dissociation" into a wider context than was intended by Wikler when he first coined it. The second conclusion by Green and Aldous is that mechanisms which reduce changes in the background EEG are different from those which affect evoked potentials and is based on their finding that while both antiacetylcholine drugs and chlorpromazine produce a similar synchronising effect on the EEG they produce different effects on primary evoked potentials in the cortex. This conclusion seems to be well-founded and is in agreement with Herz, Fraling, Neidner and Farber (1967) and White, Sewell and Rudolph (1965).

There have been relatively few attempts to demonstrate effects of antiacetylcholine drugs on electrographic recordings from sub-cortical areas despite the undoubted fact that these drugs have sub-cortical as well as cortical sites of action. This matter is discussed by White and Rudolph (1968) who devised experiments in rabbits in which electrographic recordings were obtained from the mid-brain reticular formation following extirpation of the cerebrum and diencephalon. It was possible to record spontaneous electrical activity (EEG) and potentials evoked in response to single shock stimuli applied to one sciatic nerve or to the pontine region of the reticular formation. The anticholinesterase agent physostigmine reduced or abolished the evoked potentials whereas a number of antiacetylcholine drugs had no effect upon them. However atropine, hyoscine and two psychotomimetic glycollates blocked this antagonistic effect of physostigmine on the evoked potentials, that is, they were having an action at the mid-brain level. Neither the anticholinesterase nor the antiacetylcholine agents influenced the pattern of spontaneous electrical activity recorded from the mid-brain suggesting that neurones different from those involved in evoked potentials are involved in the spontaneous activity.

The material contained in this section of the chapter is by no means exhaustive but will serve to indicate the varied types of effects on brain electrical activity which antiacetylcholine drugs have been reported to produce. It seems probable that the effects will be better understood when satisfactory methods of EEG analysis are more readily available. It is also worth making the point that in far too many of the studies single doses of single drugs have been used. The fact that these are rather specialised neuropharmacological investigations in no way detracts from the desirability of adhering to conventional pharmacological principles, that is dose-response relationships should be demonstrated if possible and preferably a number of drugs should be studied.

5.4.4 *Effects on animal behaviour*

There have been a very large number of studies on the effects of anti-acetylcholine drugs on various aspects of animal behaviour. Many different drugs, species and techniques have been used and widely differing interpretations have been put on the results. Many of these studies have involved the use of single doses of single drugs in specialised behavioural tests and it is very difficult to consider the implications of such results in any context wider than the particular experimental situation used. No attempts will be made here, therefore, to review all these investigations, rather the effects of the drugs on some aspects of behaviour will be considered with particular reference to the theme which underlies this whole chapter, that is, whether the behavioural effects of the drugs can be related to their antiacetylcholine activities or whether they are operating by mechanisms other than antagonism of acetylcholine at central cholinergic synapses. A good review of behavioural studies prior to 1966 is given by Longo (1966).

(i) *Observational methods*

There is evidence from animal behavioural studies, as well as from the human studies already referred to, of a biphasic type of action of anti-acetylcholine drugs with both excitation and depression. The type of effect produced is to some extent dose-dependent but also time-dependent. Longo (1966) pointed out the limited value of these observational studies in enabling an analysis to be made of the site and mode of action of the drugs. This is certainly correct although from most of the studies a fairly standard picture emerges with physical effects such as ataxia occurring in concert with, or sometimes preceding, signs of central nervous system depression including coma and sleep. The latter often alternate with periods of hyperactivity and apparent confusion. Various attempts have been made to liken this syndrome to other phenomena. For example, Bijlsma and Brouwer (1928) considered that it was analogous to the effects of decortication while Meyers and Abreu (1952) compared it to the delirious state which occurs in the second stage of anaesthesia.

A common source of confusion in studies of this kind results from the use of atropine and hyoscine. The well-known sedative action of the latter, which is largely lacking in the former, is undoubtedly responsible for some of the qualitatively different effects which have been reported. Many authors, however, have been unable to detect differences between the behavioural effects of the two alkaloids. For example Albanus (1970) using a double blind technique was unable to demonstrate any difference between the signs produced in dogs by atropine, hyoscine, Ditran, quinuclidine benzilate and several other piperidyl, tropinyl and quinuclidinyl esters of glycolic acid. After administration of all these drugs the dogs had stiff and stumbling gaits and difficulties with balance. They wandered around aimlessly without interest in their surroundings and often barked or whined. They walked against objects and later tried to push through obstacles. Albanus described this as a central antiacetylcholine syndrome in the dog and although he was able to evoke it with methyl atropine, very large doses were required and it seems likely that some of the drug, even in its quaternised form, had penetrated to the brain. Very similar effects of atropine and hyoscine on dogs were reported earlier by White, Nash, Westerbeke and Possanza (1961). In the same study the effects of the two drugs on the behaviour of monkeys was investigated. Doses of 1.5 mg/kg of either atropine or hyoscine given intramuscularly produced within 4 minutes, overt signs of sleep. Within 5 to 7 minutes the animals awakened and struggled violently. After this, sleep and struggling occurred intermittently and were very variable in duration. Higher doses of atropine (45 mg/kg) and of hyoscine (60 mg/kg) produced the same sequence of events. In rabbits high doses (60 mg/kg) of the two drugs produced a slight reduction in the amount of spontaneous activity.

From this study by White *et al.* it was apparent that there was a marked species difference in the type of effect produced by atropine and hyoscine but that within a given species the two drugs consistently produced a standard syndrome almost irrespective of dose providing that a certain minimal dose was exceeded. The authors considered that at this minimal dose level there was saturation of central receptors.

Another point of interest to emerge from this work was that there was no simple relationship between the behavioural changes observed and the magnitude of the changes in the EEG. Thus, in rabbits where marked EEG synchrony was produced there were relatively small effects on gross behaviour. In dogs the production of EEG synchrony was more predictable than the effects on behaviour and in monkeys the two effects seemed to be totally unrelated. On the whole it seemed that the behavioural changes induced by the drugs were correlated more with the degree of encephalisation than with the magnitude of the EEG change observed.

In view of the highly subjective nature of many of the reports on observed behaviour it might be expected that more easily quantifiable, more objective, and therefore potentially more useful, results might be obtained from studies in which the observed behavioural parameters have been relatively accurately quantified or, more particularly in which some form of instrumentation is used to monitor specific aspects of behaviour.

(ii) *Effects on social behaviour*

There have been relatively few studies on the effects of antiacetylcholine drugs on social behaviour but Van Der Poel and Remmelts (1971) analysed the effects of hyoscine and methyl hyoscine on the behaviour of the rat in a solitary or in a social situation, that is, when observed either alone or in company with another rat which had also previously been kept in isolation. Various categories of behaviour were recognised and noted in these animals. In the social situation it seemed that doses of 0.1 mg/kg or higher of hyoscine reduced the frequency of those elements of behaviour belonging to the categories of aggression (attend, approach, attack, aggressive groom, aggressive posture and bite), defensive behaviour (retreat, kick, parry and submit) and play (jump and over). Methyl hyoscine in doses of up to 10 mg/kg had no effect on the frequency of any of the behavioural patterns so that the behaviour changes presumably resulted from central actions of the drug. In the solitary situation rats showed almost exclusively exploration (explore and rearing) and maintenance (fast movements with the forepaws, wash, groom and scratch) behaviour. The former was not affected consistently by hyoscine but the drug caused complex changes in the latter. The behavioural effects noted, especially in the social situation, did not seem to be at variance with the suggestion made earlier that antiacetylcholine drugs produce an introvertive state. The results also seem to be in general agreement with the findings of Janssen, Jageneau and Niemegeers (1960) who found that tertiary atropine-like amines, including hyoscine, showed selective activity in blocking isolation-induced fighting in mice.

(iii) *Spontaneous motor activity*

It has been recognised for some time that antiacetylcholine drugs produce increases in spontaneous motor activity especially in rodents and since this is an aspect of behaviour which is relatively easy to measure, several studies have been made of this aspect of the action of the drugs. Harris (1961) using photocell activity cages found that hyoscine, over a wide dose range, produced an increase in the spontaneous activity of mice. This property was shared by atropine and several other antiacetylcholine drugs but it was not possible to relate potency of the drugs in stimulating spontaneous activity to their potency in producing mydriasis—a measure of peripheral antiacetylcholine activity. It was shown however, by using pairs of isomers including d- and l-hyoscyamine and d- and l-trihexyphenidyl that antiacetylcholine activity was necessary for the stimulatory activity. It also seemed that the effect was central in origin since quaternary salts with the exception of methyl hyoscine were inactive. Even the latter had only about one-hundredth the activity of the corresponding free base.

Abood (1968) also noted that the measurement of hyperactivity in rats is usually a good indication of central nervous system stimulation. He found that within a given series of glycollate homologues, such as the

TABLE 5.3

A comparison between the antiacetylcholine potencies of eight drugs and their potencies in increasing the level of spontaneous activity in rats

Antiacetylcholine drug	Approx. minimal effective doses (mg/kg) to produce increases in spontaneous activity in rats in photocell cages	Antagonism of oxotremorine effects in mice ED50 (mol/kg) for antagonism of oxotremorine-induced:—		Antagonism of acetylcholine on isolated guinea-pig ileum	Production of mydriasis in mice. Potency relative to atropine
		Salivation	Tremors	pA_2	
Atropine	3.13	0.44	16.2	8.4	1.0
Hyoscine	0.19	0.05	1.1	9.0	5.0
N-methyl-3-piperidyl benzilate	0.39	1.05	1.8	8.45	0.20
N-ethyl-3-piperidyl benzilate	1.56	7.1	2.7	8.1	0.12
N-methyl-3-piperidyl phenylcyclopentyl glycollate	0.78	1.2	2.0	8.4	0.39
N-ethyl-3-piperidyl phenylcyclopentyl glycollate	0.78	1.5	3.1	7.95	0.25
N-ethyl-2-pyrrolidyl methyl-phenylcyclopentyl glycollate	0.39	1.7	3.5	8.76	1.0
Ditran	1.56	0.77	0.88	8.6	0.63

piperidyl series, hyperactivity appears to be correlated with psychotomimetic potency but in some series, for example the piperazinyl glycollate esters, there is the ability to increase spontaneous activity to a marked extent but only weak psychotomimetic activity is observed.

More recently Brimblecombe and Buxton (1971) compared the potencies of eight antiacetylcholine drugs (those listed in table 5.3) in producing increases in spontaneous activity of rats in photocell cages with various measures of their antiacetylcholine potency both central and peripheral. The results are summarised in Table 5.3.

There was no statistically significant correlation between activities of the drugs in increasing spontaneous activity with any of the measures of antiacetylcholine potency. Whether it is reasonable to conclude from so few data that the increase in spontaneous activity is mediated at least in part through non-cholinergic mechanisms is perhaps doubtful but evidence now seems to be accumulating to indicate that this is the case. Certainly central antiacetylcholine activity is necessary for increase in spontaneous activity to occur but it is likely that other, as yet not defined, factors are also operating.

(iv) *Emotional behaviour*

Emotional behaviour is not easy to define but it is generally accepted that the open-field situation evokes a pattern of emotional behaviour in rodents. In this situation naive animals, usually rats, are placed in the open-field, usually a well-lit circular open-topped drum and various aspects of their behaviour are noted. Animals respond to their new environment by preening, defaecating, urinating, rearing and exploration. In a recent study Brimblecombe and Buxton (1971) studied the effects of the eight antiacetylcholine drugs listed in Table 5.3 for their effects on open-field behaviour of rats. The results summarised in Table 5.4 show that all eight drugs caused significant increases in the total amount of ambulation of the animals and that, with one exception, they caused decreases in the amount or frequency of defaecation.

The potencies of the drugs in increasing ambulation were not correlated with their potencies in increasing spontaneous activity in photocell cages suggesting that the former effect was not merely another manifestation of hyperactivity. The increase in ambulation seen in non-drugged animals in the open-field is, in any case, probably a complex response with components including emotionality and exploration. The decrease in defaecation produced by the drugs in the open-field situation might have been expected on the basis of their spasmolytic activity but the fact that their potencies in producing this effect were significantly correlated with their potencies in causing increases in ambulation seems to suggest a central component in this action. Overall, it is possible to conclude tentatively that antiacetylcholine drugs reduce emotional responsiveness, an effect akin to that already discussed in which the drugs appeared to reduce the significance of externally presented stimuli. Van Abeelen and Strijbosch (1969) used mice in what was essentially an open-field situation.

TABLE 5.4

The effects of eight antiacetylcholine drugs on open-field behaviour in rats

	Minimal effective dose (mg/kg) for	
	Decrease in defaecations	Increase in ambulation
Atropine	0.80	1.60
Hyoscine	0.20	0.05
N-methyl-3-piperidyl benzilate	0.10	0.10
N-ethyl-3-piperidyl benzilate	1.00	5.00
N-methyl-3-piperidyl phenylcyclopentyl glycollate	No effect at 10	0.05
N-ethyl-3-piperidyl phenylcyclopentyl glycollate	0.10	0.20
N-ethyl-2-pyrrolidylmethyl phenyl-cyclopentyl glycollate	0.50	0.20
Ditran	0.10	0.10

They found that hyoscine (1.25–10 mg/kg) markedly reduced exploratory behaviour in strains of mice which normally showed high scores in this respect. Physostigmine (150 μg/kg) had the same effect and it was inferred that both drugs upset the acetylcholine/acetylcholinesterase equilibrium and so disrupted transmission at cholinergic synapses thus reducing the amount of exploration normally facilitated via these synapses. The fact that hyoscine tended to increase the amount of exploration in other strains which normally gave low scores for this parameter was taken as evidence to support the hypothesis. It was suggested that in these animals the relatively too high levels of transmitter substance were blocked by the drug whereas physostigmine, which reduced exploration in these animals even more, would be expected to potentiate the effects of these postulated high levels of acetylcholine. The authors concluded from these studies that there exists in mice a cholinergic mechanism which facilitates exploratory tendencies.

(v) *Spontaneous alternation*

Rats placed in a simple T maze twice in quick succession have a high probability of showing spontaneous alternation, that is, they are likely to enter the opposite limb of the T on the second occasion. The mechanisms involved in this behaviour are still a matter for debate (Dember and Fowler, 1958) but presumably memory is involved since some recall of the direction taken on the previous run is necessary for the choice of the opposite direction in the subsequent run. Meyers and Domino (1964) found that both hyoscine (0.13 or 1.05 mg/kg) and 1-hyoscyamine (1 or 8 mg/kg) caused rats to alternate significantly less than control animals. The group given 1-hyoscyamine alternated at a chance level but the hyoscine group alternated significantly less than would be expected by chance. This result suggests that in part the drugs were producing an amnesia, an effect discussed later in this chapter, but the effect with hyoscine cannot be explained solely on this basis and it seems that the drug may have produced a tendency towards response preservation. The authors noted the similarity between these effects on behaviour and those seen in animals with hippocampal lesions and they suggested the possibility of a hippocampal site of action for the drugs.

Parkes (1965) also found that hyoscine (ED50 0.1 mg/kg) reduced alternation but only to a level approaching that which would be expected by chance. Atropine also had this effect but only seemed to have between 2 and 5 per cent the activity of hyoscine. It is of some interest in this context that Brimblecombe and Green (1968) found that atropine had about 7 per cent the central antiacetylcholine potency of hyoscine as measured by activity in blocking oxotremorine-reduced tremors.

Bureš (1968) using a slightly different form of apparatus found that doses of atropine of 6 and 15 mg/kg did not significantly impair alternation in rats whereas 0.25 and 0.5 mg/kg physostigmine had a significant effect. In another series of experiments the left-right alternation was elaborated by punishing non-alternating responses with electric shock. In this situation both 0.5 mg/kg physostigmine and 6 mg/kg atropine caused a clear impairment of the alternation habit.

Another point to be considered in connection with these experiments of Bureš is that the same animals were used throughout and so became progressively overtrained. The drugs became progressively less effective, perhaps because of this, although tolerance to the effects of the drugs must also be considered as a possibility. The other factor to be considered is that the alternation was learned in the presence of the drug and this may be an example of state-dependent or drug-dependent learning where the conditioned response is not necessarily transferred from the drugged to the non-drugged state.

Bureš noted that the results of his experiments, while showing that both atropine and physostigmine impaired the learning of new habits did not necessarily imply that the effect was due to loss of recent memory.

(vi) *Conditioned behaviour*

There are numerous reports of antiacetylcholine drugs affecting processes which are variously described as "learning" or "memory". The phenomenon mentioned in the last section of this chapter, blockade of alternation behaviour, has been ascribed, at least in part, to an amnesic effect of the drugs. That antiacetylcholine drugs do produce amnesia in human subjects is established beyond doubt, (see, for example, Innes and Nickerson, 1965) and very many studies on animals seem to demonstrate a similar phenomenon. Practically all these studies involve conditioned responses of one sort or another and since "amnesia", that is, an inability to re-call past experience, in these experimental conditions can be a manifestation of effects of drugs on acquisition, retention or recall of the conditioned response, it is essential that the precise experimental procedure should be described and it is certainly better to express the effects of the drugs specifically in terms of these rather than to use words like "learning" or "memory" which are often ill-defined and used somewhat loosely.

The results of conditioning experiments, even when essentially similar techniques are used, are exceedingly variable. Longo (1966) summarised in tabular form the results of some twenty studies of the effects of atropine and hyoscine on discrete trial avoidance using such techniques as shuttle-box and pole-jumping conditioning. In half of these studies there was no effect on performance, in three there was an improvement in performance and in the remaining seven there was a decrement in performance. Longo pointed out a few relevant facts. He noted that adverse effects tended to be observed when the conditioned response had been recently acquired or when the avoidance involved relatively complex motor performance. It is also perhaps noteworthy that in most of the experiments rats were the species used. In the one study reported by Longo, which involved monkeys, a dose of 1 mg/kg of atropine blocked the conditioned response.

Longo then summarised ten studies with atropine and hyoscine in which some element of discrimination was introduced into the avoidance procedure. Among these studies the drugs were without effect in four and produced worsening of the performance in six.

Longo also summarised six studies in which effects of the two drugs were studied on continuous avoidance, that is, level pressing in an operant conditioning situation. The results of such studies are more complex since the effects of the drugs on the number of shocks avoided can be assessed as can their effects on rate of responding. In these six studies there was a general tendency for increase in response rate; the efficiency of avoidance responding was either unaffected or reduced.

Finally, Longo tabulated seventeen studies in which rats were used most commonly but which also involved, in a few instances, dogs, rabbits or monkeys, and where there was a positive reinforcement of the conditioned response, that is, the response was to obtain a reward rather than to avoid punishment. In all cases atropine and hyoscine produced decrements in performance. There has, however, as is pointed out by Longo, been some discussion about results of this kind since antiacetylcholine drugs have central anorexic actions and produce peripheral effects including dryness of the mouth which may affect performance in food reward situations. However, some authors have controlled for this effect and it does seem that the antiacetylcholine drugs have specific effects on this type of conditioned behaviour. Another similar point worth mentioning is that these drugs produce mydriasis which may be responsible for decrement of performance in situations, especially those involving discrimination, where visual cues are used.

No very clear-cut picture emerges from these studies. Antiacetylcholine drugs commonly disrupt conditioned behaviour, often have no effect upon it, and occasionally produce increments in performance. It is not possible from this to draw any clear conclusions as to the role of cholinergic mechanisms in those aspects of behaviour involved in the various types of conditioning described above. Still less is it justifiable to extend such conclusions to the processes of learning in general. The remainder of this section will be devoted to a few specific investigations in which the effects of drugs have been studied in some detail and from which it may be possible to draw some rather more concrete conclusions.

An experiment by Bureš (1968) was designed to determine the influence of cortical mechanisms on the behaviour of rats in an avoidance situation. Both normal and functionally decorticated rats were used. A runway with a grid floor running half its length was used. Rats were conditioned to cross or escape from the grid portion of the runway within 5 seconds of being placed at the start, otherwise electric shocks were delivered via the grid. Following physostigmine treatment rats acquired the avoidance reaction slower than controls, but atropine had no effect on the rate of acquisition. Functional decortication was achieved by application of 25 per cent KCl solution to the exposed cerebral cortex. This produced a bilateral cortical spreading depression and although the rate of acquisition of the escape response was impaired animals did reach the required criterion eventually. Physostigmine had no effect on the performance of these animals but atropine produced a marked effect. Animals acquired the avoidance response as rapidly as controls but on reaching the goal box they returned spontaneously to the grid portion of the runway where they were shocked. This behaviour was seen in many animals with cortical spreading depression but was more marked in animals which had been given atropine. Bureš interprets the results as not meaning that "learning ability" was improved but that the animals avoided the shocks due to their hyperactivity. This conclusion was supported by the fact that when these animals were tested 24 hours later their performances did not differ from those of normal naive animals.

Since the stimulatory effect of atropine was enhanced in the functionally decorticated rats it seemed that the site of action in producing this effect was subcortical. On the other hand, since the behavioural impairment due to functional decortication was not modified by physostigmine, it seemed that this drug might affect those regions directly or indirectly depressed by the KCl solution.

Burěs suggested that functional decortication might produce an imbalance between the remaining functional portions of two antagonistic systems producing a cholinergic dominance which was blocked by atropine. He assumed, further, that no definite behaviour could be assumed to be specifically dependent on cholinergic transmission and that non-cholinergic systems exist which duplicate most behavioural functions of cholinergic systems. These systems might take over behavioural functions which are initially impaired.

To some considerable extent this argument by Bureš hinges on the assumption that the apparent improvement in performance of rats given atropine was due to their hyperactivity. Recently Brimblecombe and Buxton (1971) reported some results obtained in a slightly different behavioural situation, which seem to be at variance with this assumption. The same eight drugs as were listed in Table 5.4 were used and their effects studied on a conditioned avoidance response in rats. An automatic shuttlebox apparatus was used with a tone as the conditioned stimulus. All experiments consisted of two training sessions of fifty trials each, separated by an interval of twenty four hours. All the antiacetylcholine drugs were given subcutaneously 15 minutes before the first session, initially at a dose of 25 mg/kg. No drug treatment preceded the second session. The results are summarised in Table 5.5 and show that three of the drugs produced significant increases in the number of avoidances in the first session and that three gave significant decreases in avoidances during the second session. All the drugs caused significant increases in numbers of inter-trial crossings, that is, crossings of the barrier unrelated to the stimulus, during the first session and their potency in this respect was significantly correlated with their potency in causing increases in spontaneous activity measured in photocell activity cages.

TABLE 5.5

The effects of eight antiacetylcholine drugs on avoidance conditioning in rats using a shuttle-box. (Maximum dose used—25 mg/kg)

Drug	Minimal effective dose (mg/kg) for:— Increase in intertrial crossings Session 1	Increase in avoidances Session 1	Decrease in avoidances Session 2
Atropine	25.0	—	—
Hyoscine	12.5	—	—
N-methyl-3-piperidyl benzilate	0.8	0.8	6.3
N-ethyl-3-piperidyl benzilate	12.5	25.0	—
N-methyl-3-piperidyl phenylcyclopentyl glycollate	12.5	—	—
N-ethyl-3-piperidyl phenylcyclopentyl glycollate	1.6	3.1	1.6
N-ethyl-2-pyrrolidylmethylcyclopentyl	12.5	—	—
Ditran	12.5	—	25.0

The question arises as to whether this hyperactivity of the animals is solely responsible for the increases in avoidances shown in the first session following three of the drugs. The results with one of the drugs is presented in more detail in figure 5.4.

From this figure it can be seen that whereas there was a progressive increase in the number of avoidances made by the drug-treated animals during the first session, the number of inter-trial crossings showed a tendency to be reduced in numbers during the latter part of this session. If the two had been directly related they would have gone in parallel. A similar effect was seen with the other two drugs which caused similar increases in avoidances. Additionally, a check was made of the frequency of barrier crossing during the different phases of the trial period. If all the animals are assumed to cross the barrier during the 6 seconds of shock (and in practice they did) then during the remaining two periods of the trial, the intertrial period (18 seconds) and the conditioned stimulus period (6 seconds), the frequency of crossing if it is random should be in the ratio of 3:1. In fact the ratio was very much nearer 2:1 indicating a learned response to the conditioned stimulus.

It seems, therefore, that in this experiment the increased avoidance was not directly due to hyperactivity although this may have had an indirect effect in that drug-treated animals would have been provided with

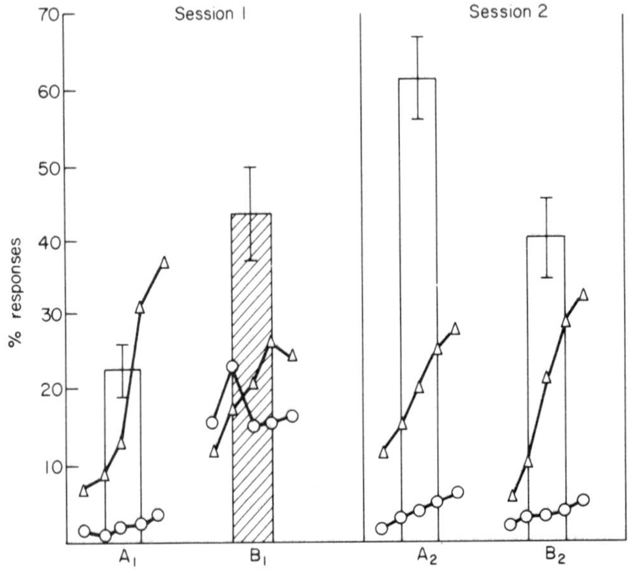

FIGURE 5.4 Effects on conditioned avoidance (shuttlebox) of N-methyl-3 piperidyl-phenylcyclopentylglycollate (25 mg/kg). Vertical blocks: mean percentage avoidance ±S.E.M. A_1 and A_2, control group; B_1 and B_2, drug group, injected 15 minutes before first session (shaded block). (△) percentage avoidances occurring during the 5 successive 10-trial periods of the daily session; (○) percentage intertrial responses occurring during the 5 successive 10-trial periods of the session.

the opportunity to make "avoidance" responses earlier than controls and so become conditioned more readily. This shows however, that the animals must have been receptive to the acquisition of the conditioned response and does not, in any case, seem to be the complete explanation since reference to Table 5.6 shows that drugs which were most effective in producing hyperactivity were not necessarily those which were most active in causing increased avoidances. The effect seems, at least in part, to be a "real" or direct effect of the antiacetylcholine drugs.

These results are in general agreement with those of Oliverio (1967) who studied the effects of hyoscine on avoidance conditioning of mice in shuttleboxes and found that an intraperitoneal dose of 2 mg/kg enhanced the performance of naive animals irrespective of whether the conditioned stimulus (light or buzzer) only preceded the unconditioned stimulus (shock) or preceded it and overlapped it. In contrast the drug markedly impaired the performance of previously trained mice, that is, those which had been given five training sessions prior to the session which followed drug administration.

Oliverio then studied the effects of hyoscine (2 mg/kg) on the responses of animals to a stimulus differing from that to which they had been trained to respond. Under these circumstances the drug again enhanced performance and it seemed that the enhancement was related to the novelty of the conditioned stimulus. Two experiments were performed to assess whether the impairment or enhancement of performance reflected actions in the processes involved in acquisition or in retention of the response. In the first experiment naive animals were given hyoscine before a session of 50 trials and showed enhanced performance compared to controls whereas when retested 24 hours later without further drug administration the control and drug-treated groups did not differ in their performance. Secondly, animals were given a training session of 50 trials and injected with either hyoscine or saline before a second session. The drug impaired performance during the five sessions following drug treatment. Twenty four hours after the last of these sessions a further session revealed no difference in performance between the two groups.

It seems from these experiments that the enhancing or impairing effects (the latter sometimes described as amnesic effects) are of short duration lasting for less than 24 hours. This part of Oliverio's findings differs somewhat from the results of Brimblecombe and Buxton (1971) referred to above. These authors, using rats, found that three out of eight antiacetylcholine drugs produced decrements in performance in a shuttlebox session 24 hours after a first session which followed immediately after drug treatment. What is more, two out of these three drugs had been responsible for enhanced performance during the first session. It is impossible to say from the evidence available whether these results obtained by Brimblecombe and Buxton are due to effects of the drugs on acquisition or retention or whether they afford another example of state-dependent learning with a lack of ability to transfer conditioning from a drugged to an undrugged state.

In his paper Oliverio remarks that hyoscine seems to exert contrasting effects depending on the type of test and on the species of animal. This is clearly true but the situation may be even more complex than this since there seems to be some indication that the effects produced may differ according to the antiacetylcholine drug which is studied. For example Bignami (1964) reported a persistent facilitation (lasting for at least six days) of avoidance conditioning in rats. This is an effect different from that reported by either Oliverio or Brimblecombe and Buxton. In Bignami's experiments a slightly different type of apparatus was used but, in addition, he was using benactyzine (see formula 117) as the antiacetylcholine drug. This drug had previously been reported by Jacobsen and Sonne (1955, 1956) to facilitate avoidance conditioning in rats in a shuttle-box situation using both partially trained animals and so-called "low-avoiders" which consistently performed badly.

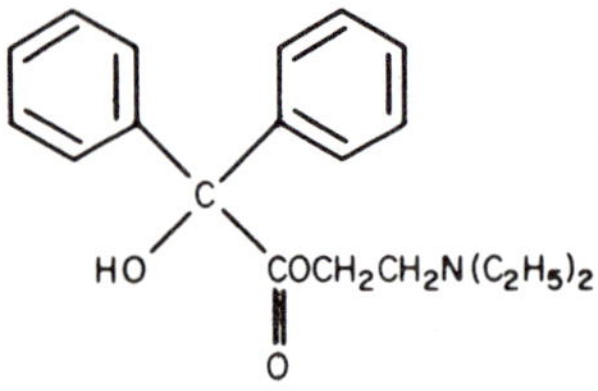

117

Despite the differences in detail between the results of Oliverio, Bignami, and Brimblecombe and Buxton they all indicate clearly that antiacetylcholine drugs can produce facilitation of avoidance conditioning in rats and mice and it would be possible to quote other examples. Many authors have, however, shown an opposite effect, that is blockade of conditioned avoidance responses. A typical study is one by Meyers, Roberts, Riciputi and Domino (1964) who used rats in a pole-jump apparatus where the animals jumped on to a pole in response to a buzzer in order to avoid an electric shock. Both low and high doses of hyoscine (0.13 and 1.05 mg/kg) and of 1-hyoscyamine (1.0 and 8.0 mg/kg) produced significant retardation in the acquisition of this response but retention of the response was virtually unaffected. The authors considered that a probable explanation of their findings was that the antiacetylcholine drugs were impairing recent memory. Since medial temporal structures, especially the hippocampus, seem to be associated with recent memory it was suggested that this might be a critical site of action of the drugs.

This example, and many other similar ones serves to show the confusing picture which exists concerning the effects of antiacetylcholine drugs on avoidance conditioning. Although the differences in apparatus and experimental procedure are often small it seems that these must account in large degree for the differing results which are obtained.

Perhaps a reasonably plausible partial explanation, is that in some experimental situations the hyperactivity produced by the drugs is able, because of the physical nature of the apparatus, to produce apparent avoidances which give the animals a better opportunity to acquire true avoidances.

The results of experiments in which the effects of antiacetylcholine drugs on operant conditioning have been studied are also somewhat confused. A fairly typical study is that by Pradhan and Roth (1968). Rats were used and were trained on three multiple schedules, fixed ratio water reinforcement alternated with time-out, shock-avoidance alternated with time-out and Sidman avoidance, time-out fixed ratio water reinforcement and time-out. The drugs used were atropine, hyoscine, Ditran, a tertiary analogue of Ditran and quaternary analogues of all these drugs.

It was found that under fixed ratio water reinforcement, responding was significantly reduced by all the tertiary drugs but very little affected by their quaternary analogues. The responses during the time-out period were also reduced. In the shock avoidance and time-out schedule Ditran, atropine and hyoscine produced a reduction in avoidance and time-out responses with increases in number of shocks received. There were however, considerable variations in the results. In the shock avoidance and time-out schedule Ditran, atropine and hyoscine produced a reduction in avoidance and time-out responses with increase in number of shocks received. In the four-phase multiple schedule a similar general pattern emerged with decrease in responding and increase in shocks. In these experiments the potencies of Ditran and hyoscine were similar with atropine being at least ten times less active. It was noted by Pradhan and Roth that since these antiacetylcholine drugs caused increases in spontaneous motor activity the decrease in learned responses could not be attributed to gross motor dysfunction. The decrease in responding in the fixed ratio water reinforcement phase of the first schedule might have been due to the fact, demonstrated by the authors, that the drugs caused a decrease in water intake. This finding is perhaps surprising since the drugs would be expected to cause dryness of the mouth.

It is of interest that the data suggested that the depressant effects of the drugs on performance became more marked as the complexity of the schedule increased and the authors considered that the drug effects might be related at least partially, to a temporary disorder in the memory process.

This matter was also studied, from a rather different point of view, by Heise and Lilie (1970). They noted that antiacetylcholine drugs increased the probability of responses occurring in situations in which their occurrence was normally much reduced due to extinction, habituation or punishment. Clearly, this phenomenon could also be explained in terms of memory deficit. It had been studied previously by Heise, Laughlin and Keller (1970) who alternated series of reinforced trials with series of non-reinforced trials. It was found that antiacetylcholine drugs impaired performance equally whether the external stimuli on trials during reinforcement periods and withdrawal periods were the same or different. The conclusion was drawn that the drugs impaired inhibitory stimulus control of responding. The follow-up study by Heise and Lilie (1970) was essentially similar except that the present and recent stimuli controlling the rat's behaviour were more explicitly defined. Rats operated a lever for sugar-water reinforcement but a trial was only reinforced if preceded by three trials during which no responses had occurred. In a first series of experiments trials were signalled by the illumination of a panel light whereas in a second series, trials preceded by three or more consecutive no-response trials, that is, trials in which lever presses would be reinforced, were signalled by the light but all other trials were signalled by the light plus a tone.

In the first series it was found that control animals responded most commonly on the trial following either two or three non-response trials whereas treatment with atropine, hyoscine or d-amphetamine produced a decrease in the number of non-response trials that preceded the occurrence of trial responses and so decreased the proportion of trial responses that produced reinforcement.

In the second series of experiments in control animals the tone stimulus clearly inhibited premature trial responding but this inhibition seemed to be removed by hyoscine and, to some extent, by atropine whereas d-amphetamine was without affect. So hyoscine impaired non-responding (that is, increased non-reinforced responding) both when the behaviour was controlled by prior events but also when it was controlled by discriminative stimuli immediately present at the time of response. Amphetamine was only effective in disrupting the former type of behaviour.

The precise implications of this result are not clear but they do, at least, suggest that the effects of antiacetylcholine drugs, at any rate in this situation, cannot be attributed solely to actions on short term memory.

The operant conditioning studies just described suffice to exemplify many investigations in which antiacetylcholine drugs have produced disrupted performance often manifested by a reduced rate of lever pressing. An example of a study in which an antiacetylcholine drug increased the rate of responding is that of Tapp (1965) who showed that the rate of lever pressing by rats was markedly increased by atropine but little affected by methyl atropine. The author concluded that this high level of operant responding was, at least in part, a function of the increased activity level produced by atropine. He stated that his results were consistent with the hypothesis that acetylcholine serves as a central transmitter substance for inhibitory neural processes.

Little purpose would be served by citing further examples of studies of the effects of antiacetylcholine drugs on operant conditioning. As with avoidance situations it is possible to find plenty of examples of both increment and decrement of performance produced by these drugs.

5.5 The role of central muscarinic mechanisms in behaviour

One of the prime objects of many pharmacological studies is to use drugs as tools to elucidate the physiological mechanisms which are being modified by the drug actions. This has certainly been the object of the majority of studies of behavioural effects of cholinergic and antiacetylcholine drugs. However, in view of the extremely variable results which have been obtained it might seem that any prospect of such elucidation is remote.

It is perhaps desirable in the first instance to attempt to find the reasons for this variability in results. There was a most penetrating analysis of the methodological problems in evaluating the role of cholinergic mechanisms in behaviour by Weiss and Heller (1969). In the first place these authors emphasise a point which has been made previously in this monograph, that dose-response relationships should be studied in behavioural, as much as in any other type of pharmacological investigation. Thus, if drugs have biphasic effects with low doses producing one effect and higher doses another effect, the choice of a single dose for study can give incomplete and sometimes misleading results. Additionally, there is the problem of extrapolation from one behavioural situation to another. Weiss and Heller quote the results of Leaf and Muller (1966) who found that 1 mg/kg hyoscine enhanced Sidman avoidance performance whereas Meyers (1965) could find no effect with the same dose of the same drug in a passive avoidance situation. Even in very closely analogous test situations, similar doses can have different effects. For example, Dilts and Berry (1967) also using a passive avoidance response found that all doses of hyoscine between 0.03 and 10 mg/kg decreased response latency in the mouse. It is impossible to give any entirely satisfactory reasons for these differences. One possible explanation might lie in differences in response to the drug between species or even between strains of the same species, in route of administration of the drug or, most likely, in the precise conditions of the experiment.

The second point raised by Weiss and Heller concerns multiple effects of drugs. It is very unusual for any chemical substance to have just one specific pharmacological effect, especially when relatively high doses are used. Reference has already been made in this monograph to the sedative action of hyoscine which sets it apart from many other antiacetylcholine drugs, and to the fact that a number of antiacetylcholine drugs have local anaesthetic activity. There is also the likelihood that differences in physico-chemical properties of drugs, even if they have identical pharmacological actions, might result in different distributions and, therefore, different overall effects.

Thirdly there is the question of selection of the aspect of behaviour for study. Weiss and Heller point out the confusion which can arise from "the tendency to select a deceptively simple aspect of behaviour as an assay and then postulate biochemical mechanisms to account for drug effects which cannot be defined precisely". One particular danger lies in the selection of behavioural tests which can only display decrements in performance—the passive avoidance response situation is one such.

Another quotation from Weiss and Heller sums up this situation admirably, "the obstacles are not insuperable. Scaling them demands mainly that we confront the complexities and problems directly, rather than fallaciously eliminating them by pretending that they do not exist".

Despite all these factors a number of workers have produced hypotheses to explain the role of cholinergic systems in behaviour and, providing they are treated with some reserve, there can be no doubt that the production and subsequent experimental testing of hypotheses is the road to progress in this, as in other, scientific fields.

One of the best examples lies in the work of Carlton (1963) who considered a number of experiments with antiacetylcholine drugs in operant conditioning situations and suggested, as a result, that these drugs led to an increase in the number of unrewarded responses due to a disinhibition of the processes which normally suppress such responses. It was speculated that a muscarinic action of brain acetylcholine was involved in the normal inhibitory process. It was also assumed that this inhibitory system antagonised the effects of a reciprocal system which normally activated behaviour; in fact when activation was produced by amphetamine the effects on operant conditioning were essentially similar to those of antiacetylcholine drugs.

In a later paper Carlton (1968) developed his hypothesis further. He considered that the effects of antiacetylcholine drugs on non-rewarded behaviour could be referred to their effects on the process of habituation, that is, the process whereby repetition of a stimulus results in the loss of the effect it produced initially. Carlton quoted the results of several experiments to support his view and it is worth describing one of them here in some detail.

Rats were given 15 minutes to explore a chamber which was novel to them. One day later they were water-deprived for one day and then returned to the chamber to which a water bottle had been added. The argument was that these animals had become habituated to all aspects of the chamber except the water bottle which they would find readily because of its novelty. This was borne out by the fact that naive water-deprived animals placed into the chamber without the benefit of the 15 minutes period of prior experience did not locate the bottle nearly as readily as the "habituated" rats. "Habituated" water-deprived rats pretreated with hyoscine, however, did not show any benefit from their previous exposure and behaved like animals which were naive to the situation and this despite the fact that on their first exploratory session they were hyperactive and indulged in abnormally high amounts of exploration of the chamber. In some way it seemed that hyoscine had attenuated the habituation process and Carlton quoted other experimental evidence to support this view.

Carlton then went on to discuss the idea, commonly expressed, that antiacetylcholine drugs interfere with the normal course of memory, that is, that their effects are not specific to the habituation process. The usual type of experimental evidence which has been presented to support this view indicates that a behaviour learned under the influence of a drug cannot be recalled later in the absence of the drug. Carlton, however, pointed out that there were three basic types of experimental situation which must be considered and which would require a broader definition of memory.

In the first experimental situation, exemplified by the spontaneous alternation studies referred to earlier, the animal is under the influence of the drug in both the pre-test and the test period, that is, in two successive runs in the T maze, and the memory hypothesis states that it does not alternate presumably because it cannot recall the direction it took in the preceding run. Carlton maintains that such an effect is satisfactorily accounted for by his habituation hypothesis.

The second basic paradigm relates to the situation where the animal is under the influence of the drug only during the test period, that is, where animals are trained before drug administration to run a maze for example. The memory hypothesis states that decrement in performance is due to the fact that animals cannot recall what occurred when they were not drugged. The habituation hypothesis, on the other hand, explains the decrement in terms of an intrusion of unrewarded responses which had previously been eliminated by habituation.

The third basic paradigm has the animal under drug in the pre-test period but not in the test period. Decrement in performance is explained by the memory hypothesis on the grounds that normal memory processes do not occur under the drug. Carlton comments that an equally acceptable explanation could be given based on stimulus-change, that is, the animal learns under one stimulus condition, the drug, and so displays a deficit under another stimulus condition, no drug. But Carlton also states that both the memory and stimulus-change hypotheses are very general and should apply to situations where the habituation hypothesis would not be applicable.

He then went on to quote ample experimental evidence for deficits in performance of animals brought about by antiacetylcholine drugs in situations where negative reinforcements in the form of an electric shock is used. However, an experiment by Carlton in which reward (sweetened milk) rather than punishment was used gave a completely different result in that hyoscine did not disrupt performance. Carlton suggested that antiacetylcholine drugs interfered with the animals' response to an "initial stress" that is a noxious stimulus to which it has not yet learned an appropriate response. Normally the initial response to shock is a general suppression of behaviour which is perhaps interfered with by the antiacetylcholine drugs in such a way that there should be some improvement in performance in the early stages of conditioning. That this can be the case is demonstrated well by the work of Oliverio (1966) and Brimblecombe and Buxton (1971) referred to in the previous section of this chapter. Decrement rather than increment in performance occurring in animals which have already learned the response can still be accounted for by the intrusion of unrewarded responses into the behaviour.

Carlton puts forward a suggestion based on speculations by one of his students, Manto, to account for this effect of hyoscine in interfering with the suppression of behaviour which normally results from shock. Emotionality in the rat is often manifested by defaecation, a response under parasympathetic control so it is suggested that the classical sympathetic reaction to stress may trigger a parasympathetic "rebound" which leads to behavioural suppression and which would of course be blocked by parasympathetic blocking agents like atropine and hyoscine. Experimental evidence for this was obtained by Manto (1967) who found that hyoscine antagonised the depression of spontaneous motor activity brought about by adrenaline.

Finally, Carlton considered the question as to which areas of the brain may be involved in the effects discussed above. He noted the striking parallel between the effects of hyoscine and lesions of the hippocampus and then quoted neurophysiological evidence which suggests that hippocampal activity is under cholinergic control. So antiacetylcholine drugs could be assumed to produce a functional lesion of the hippocampus. He also referred to experiments indicating that inhibitory functions of the ventromedial hypothalamus are mediated by cholinergic mechanisms and linked these to findings that the characteristic effects of hippocampal lesions are more marked when the lesions are in ventro-posterior areas where hypothalamic-hippocampal interactions are richer.

The ideas of Carlton have been reported at some length here partly because they represent one of the few attempts that have been made to draw various, apparently conflicting, facts together into a reasonable hypothesis and partly because the present author finds the hypothesis an attractive and convincing one especially since in his later paper Carlton seems not to suggest, as he did earlier, a system, presumably adrenergic, reciprocal to the inhibitory cholinergic system. Considerable doubts as to the presence of such a system seem to have been cast by the findings of Heise and Lilie (1970) referred to earlier, that while hyoscine impaired non-responding, irrespective of whether this was controlled internally or externally, d-amphetamine only had this effect on trials under internal control conditions.

If Carlton's view is correct, that antiacetylcholine drugs attentuate the habituation process which is presumably cholinergic, or more precisely muscarinic, in nature, it implies that the drugs are modifying processes whereby sensory information is relayed to the central nervous system or processed once it is there. It is relevant to quote, in this context, the findings of Green and Aldous (1971) who, using cats, found that anti-acetylcholine drugs reduced the significance of external stimuli. This is probably just an alternative way of stating Carlton's view but it may point to experiments which need to be carried out, in particular to explore the effects of these drugs on sensory systems.

References

Abood, L. G. (1968). The psychotomimetic glycollate esters. *Drugs Affecting the Central Nervous System.* (ed. A. Burger) Marcel Dekker, New York, pp 127–167

Abood, L. G. (1970). Stereochemical and membrane studies with the psychotomimetic glycollate esters. *Psychotomimetic Drugs.* (ed. D. H. Efron) Raven Press, New York, pp 67–74

Abood, L. G. and Meduna, L. J. (1958). Some effects of a new psychotogen in depressive states. *J. nerv. ment Dis.*, **127**, 546–551

Abood, L. G., Ostfeld, A. M. and Biel, J. H. (1958). Psychotomimetic agents. *Proc. Soc. exp. Biol. Med. N.Y.*, **97**, 483–486

Ahmed, A. and Marshall, P. B. (1962). Relationship between anti-acetylcholine and anti-tremorine activity in anti-parkinsonian and related drugs. *Br. J. Pharmac. Chemother.*, **18**, 247–254

Ahmed, A. and Taylor, N. R. W. (1959). The analysis of drug-induced tremor in mice. *Br. J. Pharmac. Chemother.*, **14**, 350–354

Albanus, L. (1970). Studies on central and peripheral effects of anti-cholinergic drugs. *FOA Report, Vol. 4, No. 4,* Research Institute of National Defence, Stockholm, pp 1–17

Anochin, P. K. (1961). The multiple ascending influences of the sub-cortical centre in the cerebral cortex. *Brain and Behaviour.* (ed. M. A. B. Brazier) *Amer. Inst. Biol. Sci.*, Washington, pp 138–170

Bebbington, A. and Brimblecombe, R. A. (1965). Muscarinic receptors in the peripheral and central nervous systems. *Adv. Drug Res.*, **2**, 143–172

Bebbington, A., Brimblecombe, R. W. and Shakeshaft, D. (1966). The central and peripheral activity of acetylenic amines related to oxo-tremorine. *Br. J. Pharmac. Chemother.*, **26**, 56–67

Bignami, F. (1964). Effects of benactyzine and adiphenine on instrumental avoidance conditioning in a shuttle box. *Psychopharmacologia (Berl.)*, **5**, 264–279

Bijlsma, U. G. and Brouwer, J. E. (1928). Die Wirkung des Skopolamins in Kombination mit Cyanid, Kohlenoxyd und Luftverdünnung. *Arch. exp. Path. Pharmak.*, **138**, 190–207

Blockus, L. E. and Everett, G. M. (1957). Tremor producing drug 1,4 dipyrrolidino-2-butyne ("tremorine"). *Fed. Proc.*, **16**, 283

Bradley, P. B. (1968). The effect of atropine and related drugs on the EEG and behaviour. *Anticholinergic Drugs and Brain Functions in Animal and Man. Progress in Brain Research,* Vol. 28 (eds. P. B. Bradley and M. Fink) Elsevier, Amsterdam, pp 3–13

Bradley, P. B. (1968). Synaptic transmission in the central nervous system and its relevance for drug action. *Int. Rev. Neurobiol.*, **11**, 1–56

Bradley, P. B. and Key, B. J. (1958). The effect of drugs on arousal responses produced by electrical stimulation of the reticular formation of the brain. *Electroenceph. clin. Neurophysiol.*, **10**, 97–110

Brimblecombe, R. W. (1970). Effects of drugs which interact with central muscarinic receptors. *Drugs and cholinergic mechanisms in the C.N.S.* (eds. E. Heilbronn and A. Winter) Research Institute of National Defence, Stockholm, pp 521–527

Brimblecombe, R. W. and Buxton, D. A. (1971). Behavioural actions of anticholinergic drugs. *Biochemical and Pharmacological Mechanisms underlying Behaviour. Progress in Brain Research,* Vol. 36 (eds. P. B. Bradley and R. W. Brimblecombe) Elsevier, Amsterdam, p 115

Brimblecombe, R. W. and Green, D. M. (1967a). Central effects of imipramine-like antidepressants in relation to their peripheral anticholinergic activity. *Int. J. Neuropharmac.*, **6**, 133–142

Brimblecombe, R. W. and Green, D. M. (1967b). Further evidence for cholinergic synapses in the mesodiencephalic activating system. *J. Physiol. (Lond.)*, **194**, 16–17P

Brimblecombe, R. W. and Green, D. M. (1968). The peripheral and central actions of some anticholinergic substances. *Int. J. Neuropharmac.*, **7**, 15–21

Brimblecombe, R. W. and Pinder, R. M. (1972). *Tremors and Tremorogenic agents*. Scientechnica, Bristol

Brimblecombe, R. W., Green, D. M., Stratton, J. A. and Thompson, P. B. J. (1970). The protective actions of some anticholinergic drugs in sarin poisoning. *Br. J. Pharmac.*, **39**, 822–830

Brimblecombe, R. W., Green, D. M., Inch, T. D. and Thompson, P. B. J. (1971). The significance of differences in the potency of enantiomers of anticholinergic drugs. *J. Pharm. Pharmac.*, **23**, 745–757

Bureš, J. (1968). The effect of physostigmine and atropine on some behavioural and electrophysiological functions in rats. *Anticholinergic Drugs and Brain Functions in Animals and Man. Progress in Brain Research*, Vol. 28 (eds. P. B. Bradley and M. Fink) Elsevier, Amsterdam, pp 61–72

Cairncross, K. D., Gershon, S. and Gust, I. D. (1963). Some aspects of the mode of action of imipramine. *J. Neuropsychiat.*, **4**, 224–231

Campbell, L. B., Hanin, I. and Jenden, D. J. (1970). Gas chromatographic evaluation of the effects of some muscarinic and antimuscarinic drugs on acetylcholine levels in rat brain. *Biochem. Pharmac.*, **19**, 2053–2058

Carlton, P. L. (1963). Cholinergic mechanisms in the control of behaviour by the brain. *Psychol. Rev.*, **70**, 19–39

Carlton, P. L. (1968). Cholinergic mechanisms in the control of behaviour. *Psychopharmacology. A Review of Progress 1957–1967*. (eds. D. H. Efron, J. O. Cole, J. Levine and J. R. Wittenborn) American College of Neuropsychopharmacology and Psychopharmacology Research Branch, Washington, pp 125–135

Chen, G. (1958). The anti-tremorine effects of some drugs as determined by Haffner's method of testing analgesia in mice. *J. Pharmac. exp. Ther.*, **124**, 73–76

Cox, B. and Hecker, S. E. (1971). Investigation of the action of oxotremorine on the guinea-pig isolated ileum preparation. *Br. J. Pharmac.*, **41**, 19–25

Cox, B. and Potkonjak, D. (1969a). The relationship between tremor and change in brain acetylcholine concentration produced by injection of tremorine or oxotremorine in the rat. *Br. J. Pharmac.*, **35**, 295–303

Cox, B. and Potkonjak, D. (1969b). The effects of atropine and dyflos on tremor and increase in whole brain acetylcholine produced by injection of oxotremorine in the rat. *Br. J. Pharmac.*, **35**, 521-529

Cox, B. and Potkonjak, D. (1969c). Effect of drugs on tremor and increase in brain acetylcholine produced by oxotremorine in the rat. *Int. J. Neuropharmac.*, **8**, 291–297

Cox, B. and Potkonjak (1970). An investigation of the tremorogenic effects of oxotremorine and tremorine after stereotaxic injection into rat brain. *Br. J. Pharmac.*, **38**, 171–180

Coyle, J. T. and Snyder, S. H. (1969). Antiparkinsonian drugs; inhibition of dopamine uptake in the corpus striatum as a possible mechanism of action. *Science*, **166**, 899–901

Cuculic, Z., Bost, H. and Himwich, H. E. (1968). An examination of a possible cholinergic link in the EEG arousal reaction. *Anticholinergic Drugs and Brain Functions in Animals and Man. Progress in Brain Research*, Vol. 28 (eds. P. B. Bradley and M. Fink) Elsevier, Amsterdam, pp 27–39

Decima, E. E. and Rand, R. E. (1965). Oxotremorine-induced tremor in the decorticated cat. *Int. J. Neuropharmac.*, **4**, 139–148

Dember, W. N. and Fowler, M. (1958). Spontaneous alternation behaviour. *Psychol. Bull.*, **55**, 412–428

Dilts, S. L. and Berry, C. A. (1967). Effects of cholinergic drugs on passive avoidance in the mouse. *J. Pharmac. exp. Ther.*, **158**, 279–285

Elliot, H. C. (1963). *Textbook of Neuroanatomy*. Lippincott, Philadelphia

English, D. C. (1962). Reintegration of effect and psychic emergence with Ditran. A clinical report and interpretation of results. *J. Neuro-psychiat.*, **3**, 304–310

Farquharson, M. E. and Johnson, R. A. (1959). Antagonism of the effects of tremorine by tropine derivatives. *Br. J. Pharmac. Chemother.*, **14**, 559–566

Farnebo, L. O., Fuxe, K., Hamberger, B. and Ljungdahl, H. (1970). Effect of some antiparkinsonian drugs on catecholamine neurones. *J. Pharm. Pharmac.*, **22**, 733–737

Fink, M. (1960). Effect of anticholinergic drugs on post convulsive electro-encephalogram and behaviour of psychiatric patients. *Electroenceph. clin. Neurophysiol.*, **12**, 359–369

Fink, M. (1968). Brain, behaviour and anticholinergic drugs. *Anti-cholinergic Drugs and Brain Functions in Animals and Man. Progress in Brain Research*, Vol. 28 (eds. P. B. Bradley and M. Fink) Elsevier, Amsterdam, pp xii-xvi

Finkelstein, B. A. (1961). Ditran, a psychotherapeutic advance: A review of one hundred and three cases. *J. Neuropsychiat.*, **2**, 144–148

George, R., Haslett, W. L. and Jenden, D. J. (1962). The central action of a metabolite of tremorine. *Life Sci.*, **8**, 361–363

Gershon, S. and Olariu, J. (1960). JB-329. A new psychotomimetic—its antagonism by tetrahydroaminoacrin and its comparison with LSD, mescaline and sernyl. *J. Neuropsychiat.*, **1**, 283–292

Giarman, N. J. and Pepeu, G. (1962). Drug-induced changes in brain acetylcholine. *Br. J. Pharmac. Chemother.*, **19**, 226–234

Giarman, N. J. and Pepeu, G. (1964). The influence of centrally acting cholinolytic drugs on brain acetylcholine levels. *Br. J. Pharmac. Chemother.*, **23**, 123–130

Giarman, N. J. and Freedman, D. X. (1965). Biochemical aspects of the actions of psychotomimetic drugs. *Pharmac. Rev.*, **17**, 1–25

Gibbs, F. A., Gibbs, E. L. and Lennox, W. G. (1937). Effect on the electroencephalogram of certain drugs which influence nervous activity. *Arch. int. Med. (Chic.)*, **60**, 154–166

Goldstein, M., Fuxe, K., Battista, A. F., Backstrom, T. and Nakatani, S. (1970). The effects of antiparkinsonian drugs on striatal dopamine. *Fed. Proc.*, **29**, 680

Green, D. M. and Aldous, F. A. B. (1971). The effects of anticholinergic drugs, chlorpromazine and LSD-25 on evoked potentials, EEG and behaviour. *Biochemical and Pharmacological Mechanisms Underlying Behaviour. Progress in Brain Research,* Vol. 36 (eds P. B. Bradley and R. W. Brimblecombe) Elsevier, Amsterdam, p 143

Grob, D., Harvey, A. M., Langworthy, O. R. and Lilenthal, J. L. (1947). Administration of di-isopropyl fluorophosphate (DFP) to man. III. Effect on the central nervous system with special reference to the electrical activity of the brain. *Bull. Johns Hopkins Hosp.*, **81**, 257–266

Harris, L. S. (1961). The effect of various anticholinergics on the spontaneous activity of mice. *Fed. Proc.*, **20**, 395

Heise, G. A. and Lilie, N. L. (1970). Effects of scopolamine, atropine and d-amphetamine on internal and external control of responding in non-reinforced trials. *Psychopharmacologia (Berl.)*, **18**, 38–49

Heise, G. A., Laughlin, N. and Keller, C. (1970). A behavioural and pharmacological analysis of reinforcement withdrawal. *Psychopharmacologia (Berl.)*, **16**, 345–368

Herz, A., Holzhauzer, H. and Teschemacher, H. (1966). Central and peripheral action of cholinomimetic drugs and their dependence on lipid solubility. *Arch. exp. Path. Pharmak.*, **253**, 280–295

Herz, A., Fraling, F., Neidner, I. and Farber, G. (1967). Pharmacologically induced alterations of cortical and subcortical evoked potentials compared with physiological changes during awake-sleep cycle in cat. *Electroenceph. clin. Neurophysiol.*, **26**, 164–176

Holmstedt, B. and Lundgren, G. (1966). Tremorogenic agents and brain acetylcholine. *Mechanisms of Release of Biogenic Amines*. (ed. U. S. van Euler, S. Rosell and B. Unvas) Pergamon Press, Oxford, pp 439–468

Holmstedt, B. and Lundgren, G. (1967). Arecoline, nicotine and related compounds. Tremorogenic activity and effect on brain acetylcholine. *Ann. N.Y. Acad. Sci.*, **142**, 126–142

Holmstedt, B., Lundgren, G., Schuberth, J. and Sundwall, A. (1965). Tremorine and oxotremorine effects on acetylcholinesterase and cholinesterase from rat brain. *Biochem. Pharmac.*, **14**, 189–191

Huxley, J., Mayr, E., Osmond, H. and Hoffer, A. (1964). Schizophrenia as a genetic morphism. *Nature (Lond.)*, **204**, 220–221

Innes, I. R. and Nickerson, M. (1965). Drugs inhibiting the action of acetylcholine on structures innervated by postganglionic parasympathetic nerves (antimuscarinic or atropinic drugs). *The Pharmacological Basis of Therapeutics* 2nd Edition (eds. L. S. Goodman and A. Gilman) Macmillan, New York, pp 521–545

Isbell, H., Rosenberg, D. E., Miner, E. J. and Logan, C. R. (1964). Tolerance and cross tolerance to scopolamine, N-ethyl-3-piperidyl benzilate (JB-318) and LSD-25. *Neuropsychopharmacology*, Vol. 3 (eds. P. B. Bradley, F. Flügel and P. Hoch) Elsevier, Amsterdam, pp 440–446

Itil, T. and Fink, M. (1968). EEG and behavioural hallucinogens with centrally active compounds. *Anticholinergic Drugs and Brain Functions in Animals and Man. Progress in Brain Research*, Vol. 28 (eds P. B. Bradley and M. Fink) Elsevier, Amsterdam, pp 149–168

Itil, T. M., Keskiner, A. and Holden, J. M. C. (1969). The use of LSD and Ditran in the treatment of therapy resistant schizophrenics. *Dis. nerv. Syst.*, **30**, 93–103

Jacobsen, E. and Sonne, E. (1955). The effect of benzilic acid diethylaminoethylester HCl (Benactyzine) on stress-induced behaviour in the rat. *Acta. pharmac. (Kbh.)*, **11**, 135–147

Jacobsen, E. and Sonne, E. (1956). The effect of Benactyzine on the conditional responses in the rat. *Acta. pharmac. (Kbh.)*, **12**, 210–320

Janssen, P. A. J., Jageneau, A. H. and Niemegeers, C. J. E. (1960). Effects of various drugs on isolation-induced fighting behaviour of male mice. *J. Pharmac. exp. Ther.*, **129**, 471–475

Jenckner, F. L. and Ward, A. (1953). Bulbar reticular formation and tremor. *Arch. Neurol. Psychiat.*, **70**, 489–502

Jenden, D. J. (1968). Testing of drugs for therapeutic potential in Parkinson's disease. *Selected Pharmacological Testing Methods.* (ed. A. Burger) Edward Arnold, London, pp 337–361

Key, B. J. (1965). Correlation of behaviour with changes in amplitude of cortical potentials evoked during habituation by auditory stimuli. *Nature (Lond.)*, **207**, 441–442

Leaf, R. C. and Muller, S. A. (1966). Effects of scopolamine on operant avoidance acquisition and retention. *Psychopharmacologia (Berl.)*, **9**, 101–109

Lechner, H. (1956). On the influence of anticholinergic drugs on the EEG of recent closed cranio-cerebral injuries. *Electroenceph. clin. Neurophysiol.*, **8**, 714–715

Longo, V. G. (1966). Behavioural and electroencephalographic effects of atropine and related compounds. *Pharmac. Rev.*, **18**, 965–996

Manto, P. H. (1967). Blockade of epinephrine-induced decrement in activity by scopolamine. *Psychonomic Science*, **7**, 203–204

Marsden, C. D. (1968). Extending the use of anticholinergic drugs in Parkinson's disease. *Third Symposium on Parkinson's Disease.* (ed. F. J. Gillingham and I. M. L. Donaldson) E. & S. Livingstone, Edinburgh, pp 195–199

Matin, M. A. (1969). Correlation between electroencephalographic (EEG) effects and acetylcholine content of the brain after cholinolytic hallucinogens. *Jap. J. Pharmac.*, **19**, 515–519

Meduna, I. J. and Abood, L. G. (1959). Studies of a new drug (Ditran) in depressive states. *J. Neuropsychiat.*, **1**, 2–10

Mellerio, F. (1964). *L'éléctroencephalographic dans les intoxications aigues.* Marson, Paris

Meyers, B. (1965). Some effects of scopolamine on a passive avoidance response in rats. *Psychopharmacologia (Berl.)*, **8**, 111–119

Meyers, B. and Domino, E. F. (1964). The effect of cholinergic blocking drugs on spontaneous alternation in rats. *Arch. int. Pharmacodyn.*, **150**, 525–529

Meyers, B., Roberts, K. H., Riciputi, R. H. and Domino, E. F. (1964). Some effects of muscarinic cholinergic blocking drugs on behaviour and the electrocorticogram. *Psychopharmacologia (Berl.)*, **5**, 289–300

Meyers, F. H. and Abreu, B. F. (1952). A comparison of the central and peripheral effects of atropine, scopolamine and some synthetic atropine-like compounds. *J. Pharmac. exp. Therap.*, **104**, 387–395

Morpurgo, C. (1967). Interaction of imipramine and desmethylimipramine with tremorine and oxotremorine in mice. *Life Sci.*, **6**, 721–731

Neubauer, H., Gershon, S. and Sundland, D. M. (1966). Differential responses to an anticholinergic psychotomimetic (Ditran) in a mixed psychiatric population. *Psychiat. Neurol. (Basel)*, **151**, 65–80

Neumeyer, J. L., Moyer, U. V., Richman, J. A., Rosenberg, F. J. and Teiger, D. G. (1967). Pharmacologically active acetylene compounds. I. Structural modifications of oxotremorine. *J. med. Chem.*, **10**, 615–620

Ochs, S. (1968). *Elements of Neurophysiology*, Wiley, New York

Olds, M. E. and Domino, E. F. (1969). Comparison of muscarinic and nicotinic cholinergic agonists on self-stimulation behaviour. *J. Pharmac. exp. Ther.*, **166**, 189–204

Oliverio, A. (1967). Contrasting effects of scopolamine on mice trained simultaneously with two different schedules of avoidance conditioning. *Psychopharmacologia (Berl.)*, **11**, 39–51

Ostfeld, A. M., Machne, X., Unna, K. R. (1960). The effects of atropine on the electroencephalogram and behaviour in man. *J. Pharmac. exp. Therap.*, **128**, 265–272

Parkes, N. W. (1965). An examination of central actions characteristic of scopolamine: Comparison of central and peripheral activity in scopolamine, atropine and some synthetic basic esters. *Psychopharmacologia (Berl.)*, **7**, 1–19

Pfeiffer, C. C., Beck, R. A. and Goldstein, L. (1967). The modification of central nervous system (CNS) function by autonomic drugs. Amine shift responses differentiate between CNS nicotinic and muscarinic effects. *Ann. N.Y. Acad. Sci.*, **142**, 181–189

Pradhan, S. N. and Dutta, S. N. (1970). Behavioural effects of arecoline in rats. *Psychopharmacologia (Berl.)*, **17**, 49–58

Pradhan, S. N. and Roth, T. (1968). Comparative behavioural effects of several anticholinergic agents in rats. *Psychopharmacologia (Berl.)*, **12**, 358–366

Rathbun, R. C. and Slater, I. H. (1963). Amitriptyline and nortriptyline as antagonists of central and peripheral cholinergic activation. *Psychopharmacologia (Berl.)*, **4**, 114–125

Rinaldi, F. and Himwich, H. E. (1955a). Alerting responses and actions of atropine and cholinergic drugs. *Arch. Neurol. Psychiat.*, **73**, 387

Rinaldi, F. and Himwich, H. E. (1955b). Cholinergic mechanism involved in function of mesodiencephalic activating system. *Arch. Neurol. Psychiat.*, **73**, 396

Schuberth, J., Sollenberg, J., Sundwall, A. and Sorbo, B. (1965). Determination of acetylcoenzyme A in brain. *J. Neurochem.*, **12**, 451–454

Schuberth, J., Sollenberg, J., Sundall, A. and Sorbo, B. (1966). Acetylcoenzyme A in brain. The effect of centrally active drugs, insulin coma and hypoxia. *J. Neurochem.*, **13**, 819–822

Slater, P. and Rogers, K. J. (1968). The effect of triethylcholine and hemicholinium-3 in tremor and brain acetylcholine. *Eur. J. Pharmac.*, **4**, 390–394

Smirnov, G. D. and Ilyutchenok, R. V. (1962). Cholinergic mechanism of cortical activation. *Sechenov. Physiol. J. USSR*, **49**, 127–139

Spencer, P. S. J. (1965). Activity of centrally acting and other drugs against tremor and hypothermia induced in mice by tremorine. *Br. J. Pharmac. Chemother.*, **25**, 442–455

Spencer, P. S. J. (1966). The antagonism of oxotremorine effects in the mouse by thymoleptics. *Life Sci.*, **5**, 1015–1023

Stein, H. H. and Cohen, J. (1970). Study of the effect *in vivo* of oxotremorine on the activity of choline acetyltransferase (choline acetylase). *Biochem. Pharmac.*, **19**, 938–939

Szerb, J. C. (1965). Average evoked potentials and cholinergic synapses in the somatosensory cortex of the cat. *Electroenceph. clin. Neurophysiol.*, **18**, 140–146

Tapp, J. T. (1965). Cholinergic mechanisms in operant responding. *J. comp. physiol. Psychol.*, **59**, 469–472

Teas, D. C. and Kiang, N. G. (1964). Evoked responses from the auditory cortex. *Exp. Neurol.*, **10**, 91–119

Ulett, G. A. and Johnson, M. W. (1957). Effect of atropine and scopolamine upon electroencephalographic changes induced by electroconvulsive therapy. *Electroenceph. clin. Neurophysiol.*, **9**, 217–224

Van Abeelen, J. H. F. and Strijbosch, H. (1969). Genotype-dependent effects of scopolamine and eserine on exploratory behaviour in mice. *Psychopharmacologia (Berl.)*, **16**, 81–88

Van Der Poel, A. M. and Remmelts, M. (1971). The effect of anticholinergics on the behaviour of the rat in a solitary and in a social situation. *Arch. int. Pharmacodyn.*, **189**, 394–396

Weiss, B. and Heller, A. (1969). Methodological problems in evaluating the role of cholinergic mechanisms in behaviour. *Fed. Proc.*, **28**, 135–146

White, R. D. and Rudolph, A. S. (1968). Neuropharmacological comparison of subcortical actions of anticholinergic compounds. *Anticholinergic Drugs and Brain Functions in Animals and Man. Progress in Brain Research.* Vol. 28 (eds. P. B. Bradley and M. Fink), Elsevier, Amsterdam, pp 14–26

White, R. P., Sewell, H. H. and Rudolph, H. A. (1965). Drug-induced dissociation between evoked reticular potentials and the EEG. *Electroenceph. clin. Neurophysiol.*, **19**, 16–24

White, R. P., Nash, C. B., Westerbeke, E. J. and Possanza, G. J. (1961). Phylogenetic comparison of central actions produced by different doses of atropine and hyoscine. *Arch. int. Pharmacodyn.*, **132**, 349–363

Wikler, A. (1952). Pharmacologic dissociation of behaviour and EEG "sleep patterns" in dogs: Morphine, N-allylnormorphine and atropine. *Proc. Soc. exp. Biol. N.Y.*, **79**, 261–265

Wikler, A. (1954). Clinical and electroencephalographic studies on the effect of mescaline, N-allylnormorphine in man. *J. nerv. ment. Dis.*, **120**, 157–175

Wikler, A. (1968). Introduction. *Anticholinergic Drugs and Brain Functions in Animals and Man. Progress in Brain Research*, Vol. 28 (eds. P. B. Bradley and M. Fink) Elsevier, Amsterdam, pp1–2

Wilson, W. P. (1961). Observations on the effect of toxic doses of atropine on the electroencephalogram in man. *J. Neuropsychiat.*, **2**, 186–190

Wilson, W. P. and Hughes, J. L. (1964). Observations on the effect of JB-329 (Ditran) on the electroencephalogram of man. *J. Neuropsychiat.*, **5**, 310–315

Yen, H. C. Y. and Day, C. A. (1965). Evaluation of anti-tremor drugs in tremor-induced rodents. *Arch. int. Pharmacodyn.*, **155**, 69–83

6 Drug Actions at Central Nicotinic Sites

In the previous chapter it was possible to discuss the actions of a variety of drugs, both agonists and antagonists, at central muscarinic sites. In comparison the discussion in this chapter will of necessity be limited to a considerable extent to a single drug, nicotine. There has been very little work with other drugs, mainly because the great majority of agonists or antagonists effective at nicotinic receptors are quaternary salts and do not, therefore, gain access to the central nervous system when administered parenterally.

6.1 Evidence for central nicotinic receptors

The main evidence for the presence of central acetylcholine receptors analogous to nicotinic receptors in the periphery, has been obtained by using the technique of microiontophoresis. The procedure is that ions are ejected into or near single central neurones from one barrel of a micropipette by the application of an electric current. Recordings can be obtained of the electrical activity of the neurone by way of another barrel of the pipette. The technique has been described and the results discussed by, for example, Curtis, Ryall and Watkins (1965) and Bradley (1968). It is apparent from these studies that acetylcholine can excite neurones in various areas of the central nervous system including the spinal cord, midbrain, thalamus, cerebral cortex and caudate nucleus. The cholinoceptive neurones can be classified as being either muscarinic or nicotinic on the basis of experiments with specific antagonists. For example, the Renshaw cells which are situated in the ventromedial portion of the ventral horn of the spinal cord are excited by acetylcholine released from the recurrent collaterals of motor axons. Acetylcholine applied iontophoretically to Renshaw cells excites them but its action is weaker than that of nicotine, carbamylcholine or tetraethylammonium. Muscarine and muscarone are only weak excitants of the neurones. Atropine and related drugs have no specific blocking action on acetylcholine-induced firing of the neurones but the nicotinic-blocking drugs dihydro-β-erythroidine, d-tubocurarine, hexamethonium and tetraethylammonium are all effective blocking agents. These criteria seem sufficient to indicate the "nicotinic" nature of the Renshaw cell synapses.

Other synapses, for example those with the pyramidal cells of the cerebral cortex, seem to be muscarinic in nature and yet others, for example certain neurones in the thalamus and in the lateral geniculate nucleus, seem to be mixed in nature being stimulated by both nicotine and muscarine.

This very brief account will suffice to indicate that it is reasonable to suppose that there are nicotinic receptors present in the central nervous system and that interaction of drugs with these receptors might be expected to produce specific effects. It is very difficult to extend these findings from single neurones to the whole animal but an attempt to do this was made by Pfeiffer, Beck and Goldstein (1967). They considered nicotine (see formula 5) and arecoline (see formula 16) to be two mood-changing drugs with predominantly nicotinic and muscarinic actions respectively. Structurally, arecoline was stated to have the typical methyl, oxygen, oxygen (M-O-O) functional groups situated at a distance of two carbon atoms from a methylated nitrogen, an arrangement analogous to that in acetylcholine itself. The relative lack of nicotinic ($\overset{+}{N}$) activity in arecoline compared with its muscarinic (M-O-O) activity was considered to be a logical derivation of the fact that the molecule possesses all the muscarinic functional groups but only a tertiary methylated nitrogen for its nicotinic functional group. Quaternisation of the tertiary nitrogen results in a one hundred fold increase in nicotinic activity. The lack of muscarinic activity in nicotine was stated to be due to the fact that it only possesses the isosteric nitrogen of the pyridine ring for its M-O-O functional group. Although like arecoline it has a methylated tertiary nitrogen in place of the $\overset{+}{N}$ functional group its predominantly nicotinic effect was attributed to the inadequate M-O-O group and the almost normal $\overset{+}{N}$ group.

The general thesis underlying this study was that most cholinergic substances have dual actions, muscarinic and nicotinic, depending on their structures and that these dual pharmacological actions are diametrically opposed in some tissues giving a balanced effect between stimulation and depression. It was pointed out that microiontophoretic studies of the type described above did not indicate whether a stimulant or depressant response in the intact animal was due to muscarinic or nicotinic effects. A specific drug enhancement response was chosen to differentiate between these effects and it was stated that neuropharmacological differences should be found between arecoline and nicotine in the following respects:—

(1) As has been found with local application of drugs to central neurones the nicotinic stimulant effect should be fleeting and not clearly dose-dependent while the muscarinic effect should be longer-lasting and dependent on dose.

(2) The stimulant effects should differ; arecoline should directly stimulate the cortex while nicotine should indirectly stimulate the cortex via the brain stem. (This conclusion is based on the different distribution of nicotinic and muscarinic neurones as determined in microiontophoretic studies).
(3) Since nicotinic effects are assumed to be determined primarily by the $\overset{+}{N}$ functional group then the effects of both nicotine and arecoline should be influenced by the addition of N-methyl or N-ethyl functional groups. Any changes obtained should be in opposite directions when the two drugs are compared.

The experimental work to test these predictions was carried out on rabbits sedated with pentobarbital. The criterion of central nervous system stimulation was the reversal of sedation determined by quantitative EEG analysis. Reversal doses 50 per cent (RD-50s) were calculated for drugs given by intravenous injection. For nicotine this was 14 μg/kg and for arecoline 9 μg/kg. As predicted the stimulant effect of the latter drug was prolonged, reliable and dose-dependent.

Trimethylamine, again in accordance with the prediction, enhanced the effect of nicotine, reducing the RD-50 to 3.4 μg/kg but decreased the effect of arecoline, increasing the RD-50 to 16 μg/kg. Triethylamine, on the other hand, blocked the central nicotinic effect, increasing the RD-50 of nicotine to 20 μg/kg. This partial blockade of $\overset{+}{N}$ effects seemed to potentiate the M-O-O effects of arecoline since its RD-50 was reduced to 3 μg/kg.

These results were described by the authors as an amine shift response and they claim that the technique is useful for differentiating between central muscarinic and nicotinic effects. The significance of the investigation was extended by the authors in experiments in which recordings of electrical activity were taken simultaneously from two brain structures of which one was always the parietal cortex. It was found that the extent to which the reticular formation affected the cortex was increased during nicotine-induced arousal in comparison with normal wakefulness while it was decreased during arecoline arousal. Nicotine was also found to re-establish the effect of the cortex upon the reticular formation to the level which is present during normal wakefulness while with arecoline the cortex was found to affect the reticular formation to a greater extent. Similar effects on cortical-subcortical relationships were obtained with the drugs in animals in which the sub-cortical electrode was sited in the hippocampus.

These studies by Pfeiffer, Beck and Goldstein have been quoted in some detail not only because they provide additional evidence for the presence of central nicotinic receptors (although the authors studiously avoid the use of this term) but also because they represent one of the few attempts to distinguish between muscarinic and nicotinic effects in terms of the response of the whole animal. Basically the conclusion is that while drugs in both classes produce cortical arousal effects, the nicotinic action is primarily on the brain stem and hippocampus while the muscarinic action is directly on the cortical neurones.

6.2 Central actions of nicotine and related substances

6.2.1. General effects

Ever since the earliest toxicological studies with nicotine there have been numerous reports of movement disorders following administration of the drug. The descriptions of the effects are varied and range from fine tremors to violent convulsions. The early literature was well summarised by Larsen, Haag and Silvette (1961) and the point is made by these authors that even the early workers were aware that the effects originated in the central nervous system. The precise site of action of the drug is however, still not clear. There is a good deal of evidence to suggest that convulsions result from excitation of the spinal cord, a conclusion reached by René in 1878. The cerebral cortex was ruled out as the site of action as early as 1885 when Hare showed that nicotine still caused convulsions in the decerebrate frog. Much later, in 1954, Yabuno administered small doses of nicotine by stereotaxic injection into various brain areas of the cat. Convulsions were only rarely and irregularly produced. Similarly Molnar, Gyorgy, Pfeiffer and Nador (1967) gave nicotine intracerebrally at doses up to the toxic level but never observed tremors and Connor, Rossi and Baker (1966) could not demonstrate tremor in conscious cats after injections of nicotine into the caudate nucleus.

The best evidence for tremor production following the administration of nicotine seems to derive from experiments in which the drug has been given parenterally and although Stern (1966) claims never to have observed tremor induced by nicotine there are a large number of reports which contradict this finding. Examples are the papers by Silvette, Hoff, Larson and Haag (1962) and Holmstedt and Lundgren (1967). The latter authors also reported that atropine, even in large doses, had no influence in nicotine-induced tremors although they can be blocked by neuromuscular blocking agents acting peripherally and by barbiturates acting centrally. (Larson, Haag and Silvette, 1961). The drug also had a self-blocking action since a dose given within 30 minutes of a previous dose did not induce tremors. Nornicotine (see formula 118) also produced tremors in the rat at a slightly higher dose than nicotine while three synthetic derivatives of nicotine (see formulae 119, 120, 121) all produced nicotine—like tremors at doses about ten times that of nicotine.

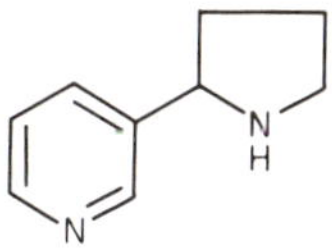

118

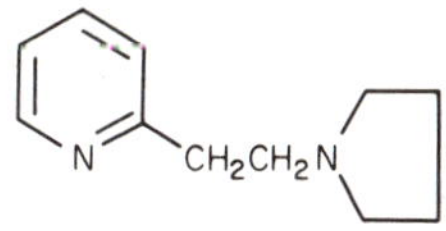

119

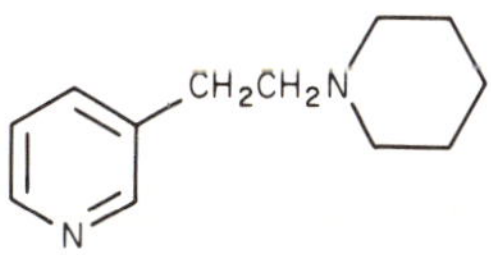

120

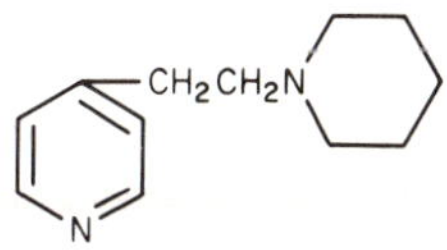

121

Two other nicotinic agents that have been studied (Molnar *et al.*, 1967) for their tremorogenic actions, this time by intracerebral administration, are tetramethylammonium bromide (see formula 122) and dimethylphenyl piperazinium iodide (see formula 123)

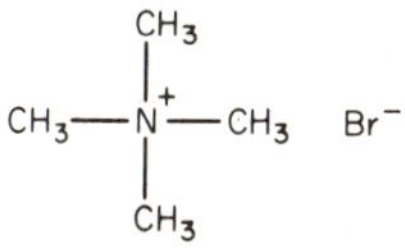

122

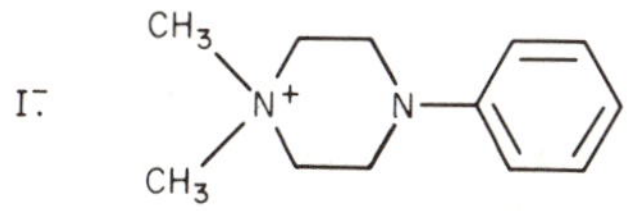

123

Rather surprisingly, unlike nicotine which was used in the same study, these agents did produce tremors in mice. These tremors were however, produced only by near-lethal doses of the compounds and were often masked by clonic convulsions.

Holmstedt and Lundgren (1967), following their previous finding that the tremor induced by muscarinic agents was accompanied by increases in brain acetylcholine levels (see Chapter 5 of this monograph), were unable to demonstrate convincing elevations in brain acetylcholine following administration of doses of nicotine sufficient to produce tremors. There are however, some reports of changes in levels of other brain amines brought about by nicotine. One such is by Westfall, Fleming, Fudger and Clark (1967) who found that tremorogenic doses of nicotine resulted in decreased overall levels of noradrenaline in the mouse brain and had specific effects in decreasing levels of dopamine and increasing levels of 5-hydroxytryptamine in the caudate nuclei of these animals. However, it seemed that the time course of these changes in brain amines did not correspond to that of the tremor so their causal relationship is perhaps doubtful.

More recently Goodman and Weiss (1973) studied the way in which nicotine affects the distribution and metabolism of ^{14}C-labelled 4-hydroxytryptamine in rat brain slices. The results indicated that nicotine, which increased 5-HT-^{14}C efflux in slices of striatum and hypothalamus, possibly did so by affecting one or more Na^{+}-dependent reactions important for 5-HT transport. This proposed, rather specific, action of nicotine might seem to be compatible with the view of Hall and Turner (1972) that nicotine-induced release of noradrenaline from brain tissue involved a more specific action than general alteration in cell membrane permeability. The significance of these findings in terms of the general central actions of nicotine is not, however, clear.

In another context there have been some suggestions that some other central effects of nicotine may result from release of acetylcholine. Armitage and Hall (1967) noted that intraventricular injection of nicotine in cats resulted in twitching of the ears, salivation, respiratory changes and a fall in blood pressure, accompanied by a bradycardia. The twitching of the ears was potentiated by the anticholinesterase drug physostigmine and it seemed possible that the effect was due to released acetylcholine. Basically the experiment was to compare the effects of carbachol and nicotine given by the intraventricular route. Carbachol produced a fall in blood pressure which was antagonised by centrally active antiacetylcholine drugs. Nicotine also produced a fall in blood pressure but it was not antagonised by these drugs. Carbachol also produced twitching of the ears, salivation and violent twitchings of the whole body. The ear response was unaffected by hyoscine but blocked by hexamethonium. Physostigmine potentiated the fall in blood pressure produced by nicotine as well as the ear response already referred to. There was no conclusive evidence that physostigmine potentiated the effects of intraventricular carbachol. Hemicholinium, which inhibits release of acetylcholine, abolished the fall in blood pressure by intraventricular nicotine and the ear response caused by nicotine or carbachol. This evidence, as is acknowledged by the authors, is by no means conclusive but does point to the possibility of at least some central actions of nicotine being mediated by acetylcholine.

The convulsions, as distinct from the tremors, produced by nicotine have also been the subjects of a number of investigations. Their central origin, as stated earlier, has been in little doubt since early work but it has been confirmed more recently by Laurence and Stacey (1952). The same authors, in 1953, suggested that the central actions of nicotine were intensified by the sympathetic discharge caused by the drug. In support of this they showed that sympathetic blocking agents reduced the convulsant activity of nicotine while this activity was potentiated by the simultaneous administration of adrenaline or noradrenaline.

There are some puzzling differences between reports of the effects of other drugs on nicotine-induced convulsions and tremors respectively. It is generally agreed that anaesthetics antagonise both manifestations of the action of the drug; Hazard and Savini (1954) considered that their site of action was mainly in the motor cortex. Similarly, neuromuscular blocking agents seem to have been found by most workers to antagonise both tremors and convulsions. The main conflict in evidence relates to atropine and related drugs which have been variously reported as having no effect or clear antagonistic effects on both tremors and convulsions. Many of these studies were summarised by Larson, Haag and Silvette (1961) and no attempt will be made here to list them. The latter authors point out, with justification, that antagonistic effects of other drugs on nicotine-induced convulsions depend on a number of factors including the species, the doses of nicotine and antagonist used, and on the precise experimental conditions. However these apparent differences between tremors and convulsions may throw some doubt on the commonly-held view that they represent a dose-related continuum of action of the drug. This doubt is perhaps intensified by the results of a careful study of the effects of 0.5 mg/kg nicotine given by intravenous injection to rats made by Orcutt, Michaelson and Prytherch (1963).

They were able to distinguish eight stages of "cerebral nicotinic effect" as follows

(1) Prostration
(2) Fine tremor
(3) Clonus
(4) Clonic convulsions
(5) Opisthotonus
(6) Tonic flexion
(7) Tonic extension
(8) Death.

A number of anti-convulsant drugs were tested for their effects against these progressive manifestations of the actions of nicotine. It was found that none of these drugs affected the nicotine prostration; only phenobarbital antagonised the tremor but caramiphen, phenacetylurea and *p*-butoxy-β-piperidyl propiophenone as well as phenobarbital antagonised the clonic convulsions. There was also a difference between the different types of convulsions since in addition to caramiphen, phenacetylurea *p*-butoxy-β-piperidyl propiophenone and phenobarbital both procaine and butacaine caused considerable reduction in the nicotine-induced tonic extension. All the antagonistic drugs mentioned reduced the lethal effects of nicotine.

Perhaps the best general conclusion to be drawn from this account is that nicotine and other nicotinic drugs produce effects due to central nervous system stimulation. Their site of action in so doing is still open to doubt and neither is it clear whether the stimulatory effects are direct ones or result from an inhibition of inhibitory systems.

This does not seem to be the whole story however for nicotine catatonia and nicotine paralysis have also been reported (see Larson, Haag and Silvette, 1961). The former is almost certainly a centrally mediated phenomenon although there could be a considerable peripheral component in the latter. It is well known that peripherally the effects of nicotine can pass from stimulation to blockade and the same seems to happen centrally. For example small doses of nicotine are emetic while large doses are not or may even be antiemetic (Larson and Silvette, 1965). Similarly, the effects on respiration are characterised by stimulation followed by depression possibly due to direct actions on the respiratory centre.

These rather generalised central actions of nicotine have been discussed at some length to indicate the complexity of its action which can clearly be influenced by factors such as species and administered dose. This complexity of action must be borne in mind in reading the remainder of this chapter which is concerned with the effects of nicotine and related drugs on the electrical activity of the brain and on specific aspects of behaviour.

6.2.2 Behavioural effects

(i) *Spontaneous activity*

The effects of nicotine on spontaneous activity exemplify very well the varied type of action the drug can have depending on the circumstances of the experiment. An early report by Nice (1913) (quoted by Larson, Haag and Silvette, 1961) showed that among 4 rats given nicotine in their drinking water and in their food, 2 showed increased activity and 2 decreased activity. There is no way of knowing, in experiments of this kind, whether the effects were related to dose. Another study in which nicotine was administered chronically, this time by injection, was made by Smith, Rosenfeld and Sacks (1935) (quoted by Larson, Haag and Silvette, 1961) who showed that the drug produced a definite decrease in activity.

Kuschinsky and Hotovy (1943) gave both single and repeated daily doses of nicotine to rats and reported marked stimulation of the animals with running movements which they stated had the same appearance as those caused by methamphetamine or caffeine. It is perhaps worthy of note that the highest daily dose used by Kuschinsky and Hotovy was 0.5 mg/kg while the dose used by Rosenfeld and Sachs was 5 to 7.5 mg/kg and it seems reasonable to conclude that, as in other situations, nicotine stimulates in small doses but depresses in high doses.

Other factors, particularly time of day, might play a significant role in the response to nicotine; Bovet, Bovet-Nitti and Oliverio (1967) used a revolving wheel for recording spontaneous activity in mice and the system was such that activity was recorded every hour during the period of light (7 a.m. to 7 p.m.) and during the period of darkness (7 p.m. to 7 a.m.). Albino rats were used and, as would be expected, their activity at night was far greater than by day. The average distance walked by day was 4 to 10 metres/hour and by night 400 metres/hour. Rats were injected with nicotine at either 10 a.m. or 10 p.m. and it was found that the drug was stimulant when injected in the morning but depressant when injected at night. Thus 0.2 mg injected at 10 a.m. produced a mean activity of 41.4 metres/hour compared to a control value of 3.5; the same dose given at 10 p.m. reduced mean activity from 408 to 139 metres/hour. Similarly a larger dose of 1 mg given in the morning increased activity from 9 to 43 metres/hour, and given in the evening reduced activity from 429 to 114 metres/hour.

It is not easy to account for these results in physiological terms. The drug seems to be levelling out the diurnal rhythm of spontaneous activity. Bovet, Bovet-Nitti and Oliverio point out that the origin of this rhythm in small mammals has not been elucidated although there is a good deal of evidence to suggest the involvement of the hypothalamic-hypophyseal axis.

To these two factors which modify the influence of nicotine on spontaneous activity, that is dose of drug and time of day, may be added three more which were listed by Morrison and Armitage (1967). The factors include strain of animal, route of administration and type of activity measured, whether exploratory or not. In experiments reported by Morrison and Armitage in the same paper two strains of mice were used, S.A.S.T.O. and Tuck's Number 1. Pairs of animals were placed in photocell activity cages and exploratory activity was allowed to subside before nicotine was injected. Doses of 0.2 mg/kg and above subcutaneously, reduced activity but no stimulant effects were seen with doses of 0.1 to 0.8 mg/kg. This non-exploratory type of activity is, presumably, not susceptible to the stimulant actions of nicotine.

Later, Morrison and Lee (1968) recorded four categories of behaviour from three strains of rats and compared the effects of nicotine and physostigmine on them. The categories of behaviour, stated to be mutually exclusive, were rearing, grooming, moving or immobile. The three strains of rats were shown to differ in their control activity with the Lister strain being most active, the Wistar strain least active and the Sprague-Dawley strain intermediate. In summary, the effects of nicotine (0.4 mg/kg) on immobility was to increase its score in rats with low control scores in this category and to decrease it in animals with high control scores. Most animals moved more after nicotine but in three animals which were particularly active in the control period moving was reduced. Rearing was reduced by nicotine in all animals except for a few with very low control scores in this category. The relationship between control score for grooming and effect of nicotine was not a clear one.

Physostigmine (0.2 mg/kg) had broadly similar effects to nicotine on all four categories of behaviour and both drugs clearly increased the activity of inactive rats and decreased the activity of active ones. This finding is precisely parallel to that of Bovet, Bovet-Nitti and Oliverio (1967) where basic activity was a function of time of day rather than of strain of animal. The similarity between the effects of nicotine and physostigmine again raises the question, referred to earlier, of whether nicotine produces many of its central effects by release of acetylcholine. These results would seem to constitute strong evidence in support of such a view. It is unfortunate that so few experiments have been carried out with other nicotinic agents. With information on additional compounds it might be possible to assess whether "nicotinic" activity is responsible for effects on spontaneous activity. One isolated report is by Mattila and Vartiainen (1962) who found that nornicotine, ethyl nornicotine and allylnornicotine, like nicotine in their experiments, all increased spontaneous activity in mice but that acetyl nornicotine and carbaminoyl-nornicotine reduced it.

(ii) *Conditioned behaviour*

Since, as was pointed out in the previous section, nicotine can have variable effects on levels of activity, there is some difficulty in interpreting the effects of the drug on various aspects of conditioned behaviour. Increased activity can give a spurious impression of facilitated performance in certain conditioning situations while decreased activity can result in non-specific decrements of performance. As might have been anticipated there are reports in the literature of both inhibition and facilitation of conditioned responses.

Early workers (for summary see Larson, Haag and Silvette, 1961), often using Pavlovian conditioning techniques, almost all reported inhibition by nicotine and it seemed to some that the drug abolished the direct relationship between strength of stimulus and intensity of response.

One of the most careful and complete studies of the effects of nicotine on conditioned behaviour was by Domino (1965) and the results are worth considering in some detail.

The first experiment was designed to study the dose-response relationships and the duration of action of nicotine on an established conditioned avoidance response in the rat. A pole-jumping technique was used in which the rats were conditioned to respond to a buzzer by jumping on to a pole in order to avoid an electric shock. In animals in which conditioned behaviour was well established doses of 0.5, and 2 mg/kg nicotine produced depression of avoidance behaviour (that is, response to the conditioned stimulus or buzzer). Escape behaviour (response to the electric shock) was not depressed so markedly. It was noted that this selective effect on avoidance behaviour was similar to that produced by tranquillisers such as chlorpromazine and reserpine. Nicotine differed from these latter drugs, however, in that large doses produced marked stimulatory effects. The author remarks that the depressant actions of nicotine on conditioned behaviour could either be secondary to post-stimulatory depression or could be due to a distinct action of nicotine. Effects on conditioned behaviour were maximal at 5 minutes after subcutaneous injection.

The effects of ganglion blocking drugs on the depressant actions of nicotine were then studied. It was found that such drugs antagonised the effects of low doses of nicotine but with large doses a peripherally acting ganglion blocker, trimethidinium, was less effective than the secondary amine ganglion blocker, mecamylamine. The conclusion seems to be that the depression of conditioned behaviour produced by low doses of nicotine has a primarily peripheral component while large doses have central as well as peripheral actions.

A further series of experiments was then carried out with monkeys which were conditioned to avoid an electric shock to their tail. The conditioned stimulus was a flickering light. The purpose of the experiment was primarily to ascertain whether a dose of nicotine which caused EEG activation had any effect on the conditioned response. In the event an intravenous dose of 40 μg/kg had no effect on the latency of this established conditioned avoidance behaviour.

Domino then turned his attention to the effects of nicotine on the acquisition of conditioned avoidance behaviour, again with the primary object of determining whether doses which produce EEG activation had any influence on this aspect of behaviour. Rats were the species used and again the pole-jump technique was used. A dose of 20 μg/kg of nicotine was without effect on the acquisition of the response; a dose of 40 μg/kg caused a slight improvement while a dose of 80 μg/kg caused a depression of acquisition and also a more rapid extinction. The facilitatory effect of 40 μg/kg could have been a direct effect on learning or could have been due to some less specific action of the drug and in an attempt to resolve this matter another experiment was carried out in which the animal was placed in a sound attentuating box and was observed through a one-way window. The effect of nicotine on the acquisition of the pole-jump in this completely isolated situation was then studied. The results were somewhat different from those of the previous experiment in which some interaction between the experimenter and the animals was possible. In the isolated situation large doses (160 and 320 μg/kg) of nicotine decreased avoidance learning but 80 μg/kg tended to facilitate learning whereas in the previous experiment it had a depressant effect. It seemed that this dose level of 80 μg/kg was a critical one and that it could produce either depressant or stimulant effects.

In a final series of experiments the effects of daily injections of nicotine on the acquisition of the pole-jump avoidance were studied. There was a tendency for the drug in daily doses of 40 and 80 μg/kg to increase avoidance learning in the rat.

These experiments by Domino indicate clearly the depressant actions of large doses of nicotine on conditioned avoidance behaviour but Domino considers that this is a manifestation of toxic actions of the drug and is irrelevant in the context of the effects of nicotine when inhaled or ingested by man. Smaller doses of nicotine had some mild facilitatory action on the acquisition of the avoidance response. This effect did not seem to be directly related to action of the drug in producing EEG arousal but Domino considered that it was not inconsistent with the hypothesis that nicotine stimulates the brain-stem activating system to produce behavioural arousal.

It is of considerable interest to compare Domino's results with those presented by Bovet (1965) at the same meeting. Avoidance conditioning was again the aspect of behaviour studied but a shuttle-box rather than a pole-jump technique was used. The conditioned stimulus was a light, the unconditioned stimulus an electric shock and the appropriate response was to cross from one half of the shuttle-box to the other. The first series of experiments was designed to study the effects of nicotine on the acquisition of the avoidance response. Subcutaneous doses of the drug were given 15 minutes before the learning trial and the first three days of the experiment. The performance of rats given 0.2 and 1.0 mg/kg was better than that of controls in that they showed significantly greater numbers of avoidance and also reached a conditioning criterion quicker. It was also noted that animals given these doses of nicotine showed greater numbers of inter-trial crossings, that is, movements from one half of the shuttle-box to the other unrelated to any particular stimulus.

This investigation was taken further by investigating the effects of nicotine on rats of a strain which had been selected on the basis of slow acquisition of avoidance conditioning and an inability to retain the response from day to day. Nicotine produced a marked reduction of the time necessary for the re-acquisition of the conditioned behaviour in these rats. It was stated that the effects of 0.2 mg/kg nicotine were comparable with those of 0.5 mg/kg amphetamine, 0.1 mg/kg lysergic acid diethylamide (LSD-25) or 5 mg/kg of the antiacetylcholine drugs benactyzine or Ditran.

Bovet (1965) found that nicotine was less active in modifying the performance of previously conditioned animals. The experimental circumstances were such that the animals successively learned two types of conditioned responses. The first was a motor response to avoid a shock upon presentation of a light stimulus and the second was inhibitory conditioning in which the animal responded to a double stimulus of both sound and light by remaining in its own half of the shuttle-box to avoid a shock. Doses of 0.2 to 1.5 mg/kg nicotine were without effect on the performance of either of these conditioned responses when they were already established. Higher doses of 3 to 5 mg/kg depressed motor conditioning and had some weak depressant action on inhibitory conditioning.

Comparison of these results with those of Domino is perhaps not particularly valuable since the experimental conditions and presumably the strains of rat used were different. The effects are basically similar with low doses tending to facilitate and high doses to depress conditioned behaviour. Bovet reached the same conclusion as Domino that the depressant effects of high doses of nicotine represented a toxic action. He considered that the stimulatory action was amphetamine-like.

As a follow up to Bovet's study, Bovet-Nitti (1965) investigated the actions of nicotine on three different or more complex forms of conditioning, maze-learning, simultaneous visual discrimination and lever-pressing avoidance. Rats were the experimental animals in each case. Using a Lashley III alley maze with either thirst or hunger as motivation she found that 0.2 mg/kg nicotine given daily before each learning trial caused a facilitation of learning especially where thirst was the motivation. The effect manifested itself as a reduction in the number of errors and as an increase in the proportion of animals which made an error-free run of the maze. These results were reported in detail by Robustelli (1963).

The effects of nicotine on visual discrimination were studied in the so-called Yerkes-Thompson Bovet-Nitti apparatus which consisted essentially of a box on the side of which were 2 to 5 doors. One of these, marked with a particular pattern, could be opened by the animal. The results were not complete at the time this paper was presented but they did indicate that nicotine could facilitate certain forms of visual discrimination.

Experiments were also carried out using a Skinner box, where animals responded to a light signal by pressing a lever in order to avoid an electric shock. Nicotine caused a reduction in the reaction time and an increase in the number of operant and conditioned responses. The number of unconditioned and inter-trial responses was, at the same time, reduced. These effects were said by Bovet-Nitti to be characteristic of nicotine and related to those of amphetamine but opposite to those of chlorpromazine.

It would be possible to quote many other examples of both facilitatory and inhibitory effects of nicotine on various aspects of behaviour. The above examples will serve to show the types of effects which have been found. In very general terms it seems that large doses of the drug tend to be depressant and small doses stimulant.

The available evidence does not, as yet, permit conclusions to be drawn concerning the precise role of central nicotinic receptors in these behavioural effects of nicotine. There is a good deal of sound evidence to suggest that the effects of the drug may be mediated, at least in part, by release of acetylcholine. If this is the case, then presumably both muscarinic and nicotinic receptors would be involved. In fact some of the actions of nicotine referred to in this chapter are reminiscent of effects of muscarinic agents discussed in the previous chapter of this monograph. Finally, the possibility that nicotine causes release of other amines, probably catecholamines, should certainly not be rejected. The findings of Bovet (1965) and Bovet-Nitti (1965) that some behavioural actions of nicotine closely resemble those of amphetamine may well be of significance in this context.

There has deliberately been no mention so far in this chapter of the effects of tobacco as distinct from nicotine. No attempt will be made here to discuss the matter in detail but reference will be made to two papers to indicate the problems involved. Basically the question is whether the central effects of tobacco smoking can be explained solely on the basis of the actions of nicotine. This matter has been discussed in detail by Larson, Haag and Silvette (1961) and Larson and Silvette (1965, 1968). The general conclusion seems to be that these effects are due to nicotine, which constitutes more than 90 per cent of the alkaloid fraction from tobacco.

However, in 1970 Bättig studied the differential effects of nicotine and the total steam volatile bases in the particulate matter of tobacco smoke on swimming endurance in the rat. The animals, with a weight on their tails equivalent to 10 per cent of their body weight, were allowed to swim to exhaustion in a tub of water. Nicotine (0.1 and 0.2 mg/kg) produced a significant improvement in performance in that there was an increase in swimming times in animals pretreated with the drug. The total alkaloid fraction, on the other hand, resulted in deterioration in performance. Nicotine treated analogously to the extraction procedure for the total alkaloid fraction has similar effects to pure nicotine. These results are not easy to explain. Clearly there could be a considerable peripheral component in the deleterious action of the total alkaloids but against that must be considered the results of Werle and Schievelbein (1961) and of Clark, Rand and Vanov (1965) who found that the alkaloids remaining in tobacco smoke after removal of nicotine had a weaker pharmacological action at a variety of peripheral sites than nicotine itself.

Later Driscoll and Bättig (1970) made a similar comparative study of the effects of nicotine and total tobacco smoke alkaloids during the extinction of avoidance behaviour in rats, that is, a purely central phenomenon. A shuttle-box was used and animals were trained to a high criterion of performance for about three months prior to the experiment. Extinction was studied by carrying out shuttle-box trials in the normal way except that the unconditioned stimulus (that is, the electric shock) was not used. Both nicotine and total alkaloids inhibited the extinction of the avoidance response to approximately the same degree, an effect described by the authors as being stimulant.

It seems, therefore, on the basis of most evidence that the central, and the majority of the peripheral effects, of tobacco smoke are due to nicotine but that certain peripheral effects may be attributable to other alkaloids.

6.2.3 Neurophysiological effects

It is clear from the work of many authors (see Silvette, Hoff, Larson and Haag, 1962) that large doses of nicotine cause typical "grand mal" seizure patterns in the EEG. Before the development of convulsions with large doses, or the main effect with smaller doses, desynchronisation of the EEG occurs. These effects can be produced by nicotine in animals transected at the mesencephalic level (Floris, Morocutti and Ayala, 1962) and although it has been suggested that the seizure discharge originates in the hippocampus there is no certainty about the site and mechanism of action of nicotine in producing EEG activation. Domino (1967) summarised the results obtained in his laboratory to that date and these warrant consideration here. Initially, an attempt was made to determine whether the effects of nicotine were direct ones or whether they were mediated through the release of, for example, adrenaline. The *cerveau isolé* preparation was used for these studies. This is a preparation in which a prepontine brain stem transection is made, the basic pattern of its EEG is indistinguishable from that seen during sleep or barbiturate anaesthesia. Small intravenous doses of adrenaline reduce EEG activation in such a preparation. In Domino's studies rabbits, cats, dogs and monkeys were used and in all of these 10 to 20 $\mu g/kg$ nicotine produced EEG activation. There was, however, some considerable species variation in the response to adrenaline and 5-hydroxytryptamine suggesting strongly that the effects of nicotine were not due to the release of these amines. Additional evidence for a direct action of nicotine on the central nervous system was obtained in dogs where all the afferent pathways to the brain were cut. Nicotine produced an EEG activation in these animals. The quaternary nicotinic drugs nicotine monomethiodide and dimethylphenyl piperazinium (DMPP; see formula 123), while producing increases in blood pressure, did not produce EEG activation. Finally, pretreatment with reserpine which depletes brain amines had no influence on the response to nicotine. All this evidence points very strongly to a direct central action of nicotine.

Attempts were then made to localise more precisely the site of action of the drug by carrying out a series of brain stem transections in the dog. The cardiovascular effects of nicotine were blocked by a quaternary ganglion blocking agent. Midpontine dogs responded to nicotine but transection at the higher, prepontine, level blocked its effect in producing EEG activation. This was not, however, the case in cats. In both species amphetamine produced EEG activation in prepontine preparations, so it seems that in dogs at any rate the systems involved in nicotine- and amphetamine-induced EEG activation are distinct with the latter being rostral to the former.

The fact that nicotine produced desynchronisation of the EEG in isolated *in vivo* reticular slabs in the dog led to the suggestion that a site of action of nicotine is on some neurones of the pontine reticular formation although, as was pointed out by Domino, this does not rule out additional sites of action. Indeed some of Domino's experiments with cats as well as the studies by Floris, Morocutti and Ayala (1962), referred to earlier, point to the hippocampus as a probable site of action.

One characteristic feature of nicotine-induced EEG activation is the rapid development of tachyphylaxis. If the drug is administered at intervals of 15 minutes to the *cerveau isolé* preparation, complete tachyphylaxis develops. This phenomenon has been used by Domino and his colleagues as a tool to distinguish between the types of central cholinergic receptors involved in the production of EEG activation. Essentially the finding was that arecoline, a drug with predominantly muscarinic actions, can still produce EEG activation in an animal where tachyphylaxis has developed to the effects of nicotine. The implication of this finding seems to be that central muscarinic receptors involved in the production of EEG activation are distinct from nicotinic receptors involved in the production of the same effect. Additional evidence for this was provided by the finding that mecamylamine (see formula 13), a secondary amine ganglion blocking agent capable of penetrating into the central nervous system, blocked the EEG effects of nicotine but not of arecoline. However, the muscarinic blocker atropine blocked both nicotine and arecoline-induced EEG activation suggesting, possibly, that the effects of nicotine are mediated finally via muscarinic receptors. Evidence in favour of this view was obtained with hemicholinium (see formula 1). This drug, which interferes with acetylcholine synthesis produces EEG changes which can be reversed by arecoline but not by nicotine. This would be expected if the effects of nicotine were due to release of acetylcholine.

In a further series of experiments Domino and his colleagues used cats with chronically implanted brain electrodes. In these animals doses of 5 to 10 μg/kg nicotine produced characteristic effects. In the initial phase, which was brief, there was desynchronisation of the neocortical EEG and considerable theta activity in the hippocampus. This was accompanied by a behavioural arousal. The animals then lapsed again into slow-wave sleep. The effects of this response of centrally and peripherally active muscarinic and nicotinic antagonists were then studied. Only the centrally acting nicotinic antagonist mecamylamine completely prevented the EEG and behavioural arousal produced by nicotine. The EEG and behavioural effects of arecoline were, on the other hand, blocked only by atropine. Atropine did, however, have an effect on the nicotine response in that it blocked the production of fast-wave sleep, the percentage of which normally increased following administration of nicotine. Mecamylamine had a similar blocking action. Again, these findings point to a specificity of effects of muscarinic and nicotinic drugs.

Domino went on in this paper to describe some behavioural effects of nicotine which were discussed in the previous section of this chapter. In summary he concluded that the initial behavioural consequences of small doses of nicotine were consistent with a brief wake-up effect, that is, an effect which might have been expected from the neuro-physiological studies.

Another approach to the problem of attempting to localise the central site of action of nicotine was made by Schaeppi (1967) who administered the drug by intravertebral, intraventricular and intracisternal injections to cats. Given by all these routes the drug produced EEG desynchronisation accompanied by retraction of the nictitating membrane, mydriasis and increase in blood pressure, effects indicating an increase in the level of vigilance of the animal. The desynchronisation of the cortical EEG was considered to be due to activation of an ascending desynchronising system resulting from an action of nicotine on the lower brain stem. These effects were usually followed by relaxation of the nictitating membrane, decrease in blood pressure and, occasionally, by EEG synchronisation. This was considered to be due to activation of an ascending synchronising system the effects of which outlasted the effects of activation of the desynchronising system.

An interesting finding was that of Stumpf and Gogolák (1967) who considered that the thyroid gland had a considerable influence on the central action of nicotine. These authors were concerned primarily with the action of nicotine on the EEG and unit activity of the hippocampus and septum in the rabbit. They found that small doses of nicotine produced EEG activation manifested by theta rhythm in the hippocampus only. Larger doses produced seizure discharges which often migrated to the septum. Septal lesions abolished the initial theta rhythm but did not affect the hippocampal seizure activity. It is of significance that the theta rhythm caused by the anticholinesterase agent physostigmine is similarly blocked by septal lesions (Mayer and Stumpf, 1958) and again this points to the possibility that nicotine acts by acetylcholine release. Treatment with the antithyroid drug sodium perchlorate tended to prevent the nicotine-induced hippocampal seizure but had little or no effect on the production of theta rhythm. The effects of sodium perchlorate were nullified if the animals were also treated with thyroxine suggesting that it was the antithyroid action of sodium perchlorate which was responsible for its effects.

It is clear, then, from the work mentioned above and from that of many other authors that nicotine produces a short-lasting behavioural and EEG arousal. This is followed by a period of EEG sedation and sleep and, often, by a period of "paradoxical" sleep, that is, behavioural sleep without a concomitant sleep pattern in the EEG. It was noted by Goldstein, Beck and Mundschenk (1967) that none of these effects seemed to be specific to nicotine, that is, they could individually be produced by other drugs although not in this particular sequence. It was possible, in the view of the authors, that mere visual examination of the EEG records was not sufficient to reveal subtle differences in the effects of nicotine as compared with those of other drugs; to investigate this possibility they recorded the electrical activity of the brains of rabbits from both cortical and subcortical (mesencephalic reticular formation, hippocampus and lateral amygdala) sites. The recorded activity was integrated to enable a quantitative analysis to be performed.

In the first series of experiments rabbits were sedated with sodium pentobarbital. A dose of 3 mg/kg intravenously produced a three-fold increase in electrical energy of activity recorded from the cortex. A dose of 10 μg/kg nicotine gave a 35 per cent decrease in electrical energy. Dose-response curves were plotted and it was found that the dose of nicotine required to produce a 50 per cent reversal of the pentobarbital effect was 14.4 μg/kg. The dose of amphetamine required to produce a similar 50 per cent reversal was 10.4 μg/kg but apart from this difference in potency there was a difference in the slopes of the dose-response regression lines for the two drugs with that for amphetamine being considerably shallower than that for nicotine. This suggests that the two drugs do not produce the final common effect by the same mechanisms. LSD-25 and psilocybin also had dose-response lines which were not parallel to that of nicotine but the line for 2-dimethylaminoethanol (deanol) was parallel to that for nicotine.

In experiments where cortical and sub-cortical electrical activities were recorded, the animals were not anaesthetised but left until spontaneous drowsiness occurred, when nicotine was administered. It was found that nicotine increased the degree to which the reticular formation influences the cortex. It shared this property with amphetamine, LSD-25 and deanol, while caffeine had the opposite effect. Cortical influence on the reticular formation was increased by caffeine and LSD-25 but amphetamine, deanol and nicotine had little effect.

Nicotine was unique among the drugs tested in maintaining cortical influence on the hippocampus and it had a tendency as did caffeine and amphetamine to increase the influence of the hippocampus on the cortex. None of the drugs seemed to reproduce the relationships which prevailed between the cortex and the amygdala during spontaneous wakefulness. In the same way varied differences and similarities between the effects of nicotine and the other drugs with regard to cortical-sub-cortical relationships were found in the stages of sedation and paradoxical sleep which succeeded the nicotine-induced arousal.

This study seemed to demonstrate that none of the other drugs chosen had completely identical actions to that of nicotine. The authors considered that the results did not allow conclusions to be drawn concerning the precise site or mode of action of the drug. It would clearly be of interest to see whether, for example, anticholinesterase agents or other nicotinic agonists mimic the actions of nicotine using this technique. This should provide strong evidence for nicotine acting by release of acetylcholine.

As in behavioural studies it is tempting to try to relate the effects of nicotine on brain electrical activity to the effects of tobacco smoking. Armitage, Hall and Morrison (1968) carried out certain studies which they considered provided an objective pharmacological explanation for the tobacco smoking habit. One finding was that doses of 2 μg/kg of nicotine given intravenously every 30 seconds for 20 minutes to an anaesthetised cat soon produced EEG activation accompanied by an increased release of acetylcholine from the cortex. The same doses of nicotine also produce EEG and behavioural arousal in unanaesthetised cats, either the *encéphale isolé* preparation or the conscious unrestrained cat. It was considered, from other pharmacological studies that forty intravenous injections of nicotine given every 30 seconds for 20 minutes represent roughly the nicotine intake resulting from smoking two cigarettes and inhaling deeply.

Later Armitage and Hall (1968) reported that tobacco smoke given by inhalation caused cortical activation and behavioural arousal in the cat *encéphale isolé*. These effects they considered to be more relevant to the pharmacology of tobacco smoking than those reported by Phillis and York (1968) who found that nicotine and tobacco smoke, as well as acetylcholine given to anaesthetised cats depressed the firing of cortical neurones. Armitage and Hall pointed out that the doses of nicotine (100 to 1000 μg/kg intravenously or 1 to 20 μg intra-arterially) are likely to be far in excess of levels achieved during cigarette smoking.

Before any realistic assessment can be made of this kind in terms of the effect on man of smoking tobacco it is necessary to know whether analagous electrophysiological effects of nicotine or tobacco smoke can be found in humans. One of the few attempts to investigate this using quantitative methods was made by Murphree, Pfeiffer and Price (1967). Their paper takes the form of a progress report of exploratory work but in general it seemed, by analogy with the effects of other drugs, that the effects of smoking on the EEG seemed to be stimulant in that there tended to be a reduction in its electrical energy.

6.2.4 General conclusions

It is clear that nicotine has biphasic actions on behaviour with small doses having stimulant actions and higher doses being depressant. The sites of action of the drug are not yet entirely clear although sub-cortical structures, particularly the reticular formation and the hippocampus are certainly involved. A direct action on cortical neurones also seems certain, at least with higher doses.

It seems very likely that nicotine produces its effects, at least in part, by release of acetylcholine. Direct actions and actions mediated via the release of other amines cannot be ruled out as additional or alternative modes of action.

The central effects of tobacco smoking seem to be almost entirely due to nicotine and these effects appear to be stimulant in nature.

The main requirement in this field is for studies with nicotinic agents other than nicotine itself. It sometimes appears that nicotine is the least typical of all nicotinic drugs and it is dangerous to generalise about central nicotinic cholinergic mechanisms on the basis of studies with this single drug.

References

Armitage, A. K. and Hall, G. H. (1967). Further evidence relating to the mode of action of nicotine in the central nervous system. *Nature (Lond.)*, **214**, 977–979

Armitage, A. J. and Hall, G. H. (1968). Nicotine, smoking and cortical activation. *Nature (Lond.)*, **219**, 1179–1180

Armitage, A. K., Hall, G. H. and Morrison, C. F. (1968). Pharmacological basis for the tobacco smoking habit. *Nature (Lond.)*, **217**, 331–334

Battig, K. (1970). Differential effects of nicotine and tobacco smoke alkaloids on swimming endurance in the rat. *Psychopharmacologia (Berl.)*, **18**, 300–304

Bovet, D. (1965). Action of nicotine on conditioned behaviour in naive and pretrained rats. *Tobacco Alkaloids and Related Compounds.* (ed. U. S. von Euler) Pergamon Press, Oxford, pp 125–136

Bovet, D., Bovet-Nitti, F. and Oliverio, A. (1967). Action of nicotine on spontaneous and acquired behaviour in rats and mice. *Ann. N.Y. Acad. Sci.*, **142**, 261–267

Bovet-Nitti, F. (1965). Action of nicotine on conditioned behaviour in naive and pretrained rate II. Complex forms of acquired behaviour. *Tobacco Alkaloids and Related Compounds.* (ed. U. S. von Euler) Pergamon Press, Oxford, pp 137–143

Bradley, P. B. (1968). Synaptic transmission on the central nervous system and its relevance for drug action. *Int. Rev. Neurobiol.*, **11**, 1–56

Clark, M. S. G., Rand, M. J. and Vanov, S. (1965). Comparison of pharmacological activity of nicotine and related alkaloids occurring in cigarette smoke. *Arch. int. Pharmacodyn.*, **156**, 363–379

Connor, J. D., Rossi, G. V. and Baker, W. W. (1966). Analysis of the tremor induced by injection of cholinergic agents into the caudate nucleus. *Int. J. Neuropharmac.*, **5**, 207–216

Curtis, D. R., Ryall, R. W. and Watkins, J. C. (1965). Cholinergic transmission in the mammalian central nervous system. *Pharmacology of Cholinergic and Adrenergic Transmission.* (eds. G. B. Koelle, W. W. Douglas and A. Carlson) Pergamon Press, Oxford, pp 137–145

Domino, E. F. (1965). Some behavioural actions of nicotine. *Tobacco Alkaloids and Related Compounds.* (ed. U. S. von Euler) Pergamon Press, Oxford, pp 145–166

Domino, E. F. (1967). Electroencephalographic and behavioural arousal effects of small doses of nicotine: A neuropharmacological study. *Ann. N.Y. Acad. Sci.,* **142,** 216–244

Driscoll, P. and Battig, K. (1970). The effect of nicotine and total alkaloids extracted from cigarette smoke on avoidance behaviour in rats under extinction procedure. *Psychopharmacologia (Berl.)*, **18**, 305–313

Floris, V., Morocutti, G. and Ayala, G. F. (1962). Azione della nicotina sulla attivatá bioelectrica della corteccia del talamo e dell' ippocampo nil coniglio. Sua azione di "arousal" e convulsivante primitiva sulle strutture ippocampotalamiche. *Boll. Soc. Ital. Biol. Sper.*, **38**, 407–410

Goldstein, L., Beck, R. A. and Mundschenk, D. L. (1967). Effects of nicotine upon cortical and sub-cortical electrical activity of the rabbit brain: Quantitative analysis. *Ann. N.Y. Acad. Sci.,* **142**, 170–180

Goodman, F. R. and Weiss, G. B. (1973). Alteration of 5-hydroxytryptamine-^{14}C efflux by nicotine in rat brain area slices. *Neuropharmac.,* **12**, 955–965

Hall, G. H. and Turner, D. M. (1972). Effects of nicotine on the release of ^{3}H-noradrenaline from the hypothalamus. *Biochem. Pharmac.,* **21**, 1829–1838

Hare, H. A. (1885). The physiological and pathological effects of the use of tobacco. *Fiske Fund Prize Dissertation No. 34.* P. Blakiston, Son & Co., Philadelphia

Hazard, R. and Savini, E. (1954). Empêchement par certains anésthesiques généraux des actions toxique, convulsivante et vasomotrice de la nicotine chez le rat. *Arch. int. pharmacodyn.,* **98,** 411–420

Holmstedt, B. and Lundgren, G. (1967). Arecoline, nicotine and related compounds. Tremorogenic activity and effect upon brain acetylcholine. *Ann. N.Y. Acad. Sci.,* **142**, 126–142

Kuschinsky, G. and Hotovy, R. (1943). Über die zentral erregende Wirkung des Nicotins. *Klin. Wschr.,* **22**, 649–650

Larson, P. S. and Silvette, H. (1965). Tobacco alkaloids and central nervous system functions. *Tobacco Alkaloids and Related Compounds.* (ed. U. S. von Euler) Pergamon Press, Oxford, pp 105–124

Larson, P. S. and Silvette, H. (1968). *Tobacco, Supplement 1.* The Williams & Wilkins Co. Ltd., Baltimore

Larson, P. S., Haag, H. B. and Silvette, H. (1961). *Tobacco: Experimental and Clinical Studies.* The Williams & Wilkins Co. Ltd., Baltimore

Laurence, D. R. and Stacey, R. S. (1952). The effect of methonium compounds on nicotine convulsions. *Br. J. Pharmac. Chemother.*, **7**, 80–84

Laurence, D. R. and Stacey, R. S. (1953). Mechanism of the prevention of nicotine convulsions by hexamethonium and by adrenaline blocking agents. *Br. J. Pharmac. Chemother.*, **8**, 62–65

Mattila, M. and Vartiainen, A. (1962). The lethal dose, pressor effect and intestinal activity of some pyrrolidine N-substituted nor-nicotine derivatives. *Acta. Pharm. Tox. (Kbh.)*, **19**, 330–336

Mayer, C. and Stumpf, C. (1958). Die Physostigminwirkung auf die Hippocampustätigkeit nach Septumläsionen. *Arch. exp. Path. Pharmak.*, **234**, 490–500

Molnar, J., Gyorgy, L., Pfeiffer, A. K. and Nador, K. (1967). Tremorogenic and convulsant actions of intracerebrally administered vegative poisons in the mouse. *Acta. Physiol. Acad. Sci. Hung.*, **31**, 249–256

Morrison, C. F. and Armitage, A. K. (1967). Effects of nicotine upon the free operant behaviour of rats and spontaneous motor activity of mice. *Ann. N.Y. Acad. Sci.*, **142**, 268–276

Morrison, C. F. and Lee, P. N. (1968). A comparison of the effects of nicotine and physostigmine on a measure of activity in the rat. *Psychopharmacologia (Berl.)*, **13**, 210–221

Murphree, H. B., Pfeiffer, C. C. and Price, L. M. (1967). Electroencephalographic changes in man following smoking. *Ann. N.Y. Acad. Sci.*, **142**, 245–260

Orcutt, J. A., Michaelson, S. M. and Prytherch, J. P. (1963). The inhibition of nicotine-induced convulsions in the rat. *Arch. Int. Pharmacodyn.*, **146**, 238–244

Pfeiffer, C. C., Beck, R. A. and Goldstein, L. (1967). The modification of central nervous system (CNS) function by autonomic drugs. Amine shift responses differentiate between CNS nicotinic and muscarinic effects. *Ann. N.Y. Acad. Sci.*, **142**, 181–189

Phillis, J. W. and York, D. H. (1968). Nicotine, smoking and cortical inhibition. *Nature (Lond.)*, **219**, 89–91

René, A. (1878). A propos de l'immunité bêtes à cornes pour la nicotine. *Gas. hôp.*, **51**, 806

Robustelli, F. (1963). Azione delle nicotina sull' apprendiments del natto nel labirints. *Rend. Accad. Naz. Lincei. Cl. Sc. firs (VIII)*, **34**, 703–709

Schaeppi, V. (1967). Effects of nicotine administration to the cat's lower brain stem upon electroencephalogram and autonomic system. *Ann. N.Y. Acad. Sci.*, **142**, 40–49

Silvette, H., Hoff, E. C., Larson, P. S. and Haag, H. B. (1962). The actions of nicotine on central nervous system functions. *Pharmac. Rev.*, **14**, 137–173

Stern, P. (1966). Methods for evaluating antitremorogenic drugs. *Methods in Drug Evaluation.* (eds. P. Mantegazza and F. Piccinini) North Holland, Amsterdam, pp 263–269

Stumpf, C. and Gogolák, G. (1967). Actions of nicotine upon the limbic system. *Ann. N.Y. Acad. Sci.*, **142**, 143–158

Werle, E. and Schievelbein, H. (1961). Tabellen zur Toxikologie, Pharmakologie und Chemie von Inhaltstoffen des Tabaks und des Tabakrauches. *Arzneimittel Forsch.*, **11**, 1011–1016

Westfall, T. C., Fleming, R. M., Fudger, M. F. and Clark, W. G. (1967). Effect of nicotine and related substances upon amine levels in the brain. *Ann. N.Y. Acad. Sci.*, **142**, 83–100

Yabuno, S. (1954). Coma puncture by means of nicotinisation. *Acta. Schlae Med. Univ. Kioto*, **32**, 32–51

7 General Conclusions

The ideal conclusion to this monograph would have been the emergence of a hypothesis postulating a precise role for central cholinergic mechanisms in controlling specific aspects of behaviour. The pharmacologist always hopes, and often believes, that by using drugs as tools he should be able to elucidate the underlying physiological systems which are being modified by drug actions. In some instances these fond hopes have been realised but even a casual perusal of this monograph makes it clear that it would be premature at the present state of knowledge to attempt to explain the role of central cholinergic systems in anything more than the most general terms. There are far too many apparently conflicting pieces of evidence and far too many results which are capable of alternative explanations to make any but the most provisional postulate quite unjustified.

This somewhat pessimistic conclusion should not be interpreted as indicating the futility of further work. On the contrary, the field is potentially a most fruitful one. Clearly cholinergic transmission occurs in the brain, or if it does not, then in Bernheim's (1966) words, the brain "is carrying a lot of excess baggage". Equally clearly this cholinergic transmission can be disrupted by a variety of drugs with consequent modification of the behaviour of the organism as a whole. There is, however, a great deal to be discovered about the system and events which link the drug action to the behavioural manifestations. It has become increasingly possible to explain, or at any rate to suggest explanations for, drug actions at the receptor in mathematical and physicochemical terms but this has, in a sense, made the drug action more remote from behaviour which is still expressed in largely empirical and descriptive terms.

It is not at all easy to see which type of investigation is most likely to provide the link between these two phenomena. Anatomical and histochemical studies designed to map central cholinergic pathways more precisely will certainly be rewarding. Additionally the pharmacologist has a great deal more work to do in attempting to produce some order out of the somewhat chaotic picture which has so far emerged from studies with drugs.

It may be of some value to explore possible reasons for the confused results emanating from pharmacological studies. Clearly the inherent complexity of the system being studied is to a large extent responsible but in many instances the experimental design and strategy leaves a good deal to be desired. All too often experiments involve the use of small numbers of animals and single doses of single drugs. Often the complexity and duration of the experimental procedure makes this almost inevitable but there is no justification for the sweepingly general conclusions which are often drawn from such limited studies. A very similar point was made by Russell (1969) in a discussion of the behavioural aspects of cholinergic transmission where he noted the logical inaccuracy of generalising from measures of behaviour in specific test situations to behaviour as a whole. Russell also stated that in the search for relations between the cholinergic system and behaviour "no single *experimentum crucis* has yet been designed. Rather, attempts to describe a particular relation have depended on the convergence of results from a number of related experiments."

Certainly one of the most important reasons for much of the confusion resulting from pharmacological studies of behaviour is due to the rather naive assumption that drugs exert only a single, specific pharmacological action. In fact most drugs have a number of actions; several examples are given by Karczmar (1969) who points out the error in interpretation of results may be so great as to lead to the postulation of a cholinergic synapse where none exists. He cites the case of hemicholinium which inhibits acetylcholine synthesis and which may be expected to depress transmission by virtue of this action alone. However, hemicholinium may also depress transmission by entirely non-cholinergic actions.

Karczmar (1969) also elaborated on the point made above concerning the use of single doses of drugs. He noted that some drugs have biphasic actions and in particular cited results with high and low doses of atropine which could be shown to produce opposite behavioural actions in rodents.

Another major difficulty is that in a system so complex as the central nervous system it is unlikely that neural pathways, particularly those involved in the more complex aspects of behaviour, will be cholinergic throughout. Thus, drugs which modify cholinergic activity may indirectly modify activities mediated by biogenic amines or other central transmitter substances. Attempts to correlate cholinergic or antiacetylcholine potency with potency in modifying behaviour might, therefore, be doomed to failure because the system is not sufficiently simple to allow such correlations to emerge.

Perhaps the greatest need is for more drugs which exert central actions when administered parenterally. The discoveries of the glycollates, an active series of centrally acting antiacetylcholine drugs, and of oxotremorine, a centrally acting muscarinic agent, have certainiy led to a better understanding of central muscarinic mechanisms. Similar improvements in the understanding of nicotinic mechanisms may well result from the development of agonists and antagonists acting at central nicotinic receptors.

References

Bernheim, F. (1966). The role of acetylcholine in the central nervous system. *J. Neurosurg.*, **24**, 168–169

Karczmar, A. G. (1969). Is the central cholinergic system overexploited? *Fed. Proc.*, **28**, 147–157

Russell, R. W. (1969). Behavioural aspects of cholinergic transmission. *Fed. Proc.*, **28**, 121–131

INDEX

A

Acetic acid 2, 4
Acetylcholine
—accumulation 66
Acetylcholine
—at autonomic ganglia 45–47
—brain levels 135
and heart 1, 2
muscarinic activity 1
nicotinic activity 1, 9, 45, 190
—structure 20, 32, 57
Acetycholine homologues
—action on autonomic ganglion 52
—action on neuromuscular junction 51
Acetylcholine receptors
—in autonomic ganglia 52
—at neuromuscular junction 52
Acetylcholinesterase
—action on acetylcholine 65
—in autonomic ganglia 45
Acetylcholinesterase inhibition 63, 72
Acetylcoenzyme A, in brain 135
Acetylenic amines
—muscarinic activity 26, 33, 133
Acetylenic carbinols
—antiacetylcholine activity 38–39
Acetylenic glycollates
—antiacetylcholine activity 38–39
Acetyl-α-methylcholine
—structure 23
Acetyl-β-methycholine, *see* Methacholine
Adrenaline
—action on postsynaptic potentials 45
—effect on EEG 205
—reduction in heart rate by 2
Affinity 3, 34
Affinity constant 31, 35
Ageing 86
Agonist/antagonist relationship 32–39
—structure and activity 37–39
Alkyl trimethylammonium salts 21
Amanita muscaria 10
Ambenonium
—structure 89
—use 5, 89
Amines
—poisoning by 119
Amino-acid amides
—nicotinic activity 55
Amino-acid esters
—ganglion-stimulating effects 46
—nicotinic activity 54
Amiton
—toxicity 74
Amitriptyline
—as muscarinic antagonist 150
Amobarbital
—use 87
Amphetamine
—behavioural effects 110, 175, 202
—effect on EEG 205, 208
—as muscarinic antagonist 141
Antagonist activity 31
—stereochemical requirements 35–37
Antiacetylcholine activity
—steric factors 36
Antiacetylcholine drugs
—assay 30
—behavioural effects 158, 160, 176–180
—behavioural effects in man 147–151
—biphasic action 160
—central effects 144–176
—combination with oxime 97
—effect on EEG 151–159
—effect on respiration 89
—neuromuscular effects 47
—and receptors 38
—stereochemical requirements of activity 35
—therapeutic use 32, 89–93
Anticholinergic drugs, *see* Antiacetylcholine drugs
Anticholinesterases
—assay 63
—behavioural effects 98–121
—central activity 63, 66–69
—effect on EEG 67, 98
—effect on respiration 13, 67–68
—muscarinic activity 66
—nicotinic activity 67, 68

Anticholinesterases (continued)
—poisoning by 11, 66–69, 71–98
—therapeutic use 63
Anticonvulants
—action 196
—use 95
Antimuscarinics, *see* Antiacetylcholine drugs
Areca catechu 10
Arecoline
—effect on behaviour 142–3
—effect on brain levels of acetylcholine 135
—effect on EEG 206
—effect on movement 141
—muscarinic activity 135, 191
—structure 10, 191
Atropa belladonna 8
—poisoning by 11
Atropine
—behavioural effects 110–111, 160, 166–170, 174–5
—effect on brain levels of acetycholine 135
—effect on EEG 112, 152, 155, 157, 161, 206
—effect on heart rate 2
—effect on movement 162, 196
—effect on neuromuscular junction 47
—effect on postsynaptic potentials 45
—effect on respiratory depression 11
—inhibition on dopamine uptake 140
—as muscarinic antagonist 1, 32, 102, 141
—poisoning 11, 147
—racemisation 32
—structure 8, 32
—use 70, 74, 87, 89, 91, 94, 97, 98, 102, 103, 114
Atropine methyl bromide 133
Autonomic ganglia
—anticholinesterases at 67
—drug actions on 1, 7, 8, 44, 45

B

Barbituric acid derivatives
—behavioural effects 111
Belladonna alkaloids
—for Parkinsonism 137
—poisoning by 11, 147
Benactyzine
—effect on behaviour 173, 202
Benzhexol
—as muscarinic antagonist 141
Benzilic acid derivatives 33
Benzoquinonium
—use 74, 90
Benzoylcholine esters
—nicotinic activity 57
Benztropine
—inhibition of catecholamine uptake 139, 140
—as muscarinic antagonist 141
—structure 139
Betel nut 10
Bethanechol
—structure 19, 38
—therapeutic use 19
Bis (dimethylamido) fluorophosphate *see* Dimefox
Bis-(4-formyl N-methylpyridinium oxime) *see* toxogonin
1, 3-bis-pyridinium-4-aldoxime) propane dibromide
—structure 78
—use 84, 88
Bis-quaternary pyridine aldoximes
—structure 78
Blood-brain barrier 10
Botulinum toxin 3, 68
M-Bromophenyl choline ether
—nicotinic activity 53
—structure 53
Butacaine
—as nicotinic antagonist 197
Butrylcholinesterase
—inhibition 70
Butyramides
—nicotinic activity 55
Butyrates
—nicotinic activity 55

C

Caffeine
—effect on LSD 208
—effect on muscle 50
Calabar 5
Calcium 144
Caramiphen
—effect on muscle 48
—as muscarinic antagonist 141
—as nicotinic antagonist 196
—structure 48
—use 91, 93, 101
Carbachol
—action 190, 195
—ganglion stimulant action 4
—at pre-synaptic terminals 46
—structure 20
—therapeutic uses 19, 76

Carbamates
—action 5, 63, 68
—poisoning by 72, 119
—use 63, 72, 74
Carbamylcholine, *see* Carbachol
Carbinols 38
Chlorpromazine
—effects on EEG 159
—as muscarinic antagonist 141
—potentiation of organophosphates 96
—use 95
Cholinacetylase, in brain 135
Choline
—effect on brain levels of acetycholine 136
—use 2, 4, 77
Choline phenyl ether 55
Cholinoceptive sites 10
—central 190–193
—presynaptic ganglionic 46
Curare 12
—neuromuscular blockade 6, 8, 50, 68
Cyclic amidines
—nicotinic activity 54, 56
Cyclic ethers 21, 22
2-cyclohexyl-2-hydroxy-2-phenylacetic acid
—esters in antagonism to acetylcholine 37

D

DAM, *see* Diacetylmonoxime
Datura alba 155
DDT
—poisoning 114
Deanol
—effect on EEG 208
Decamethonium
—neuromuscular blocking activity 51
—structure 7, 51
Demeton-S
—behavioural effects 101
—inhibition of cholinesterase 99
—structure 99
Desmethylimipramine
—as muscarinic antagonist 150
DFP, *see* Dyflos
Diacetylmonoxime
—structure 78
—use 80–83, 85
Dieldrin
—poisoning by 114
Diethazine
—inhibition of catecholamine uptake 140
—structure 140
N-(4-diethylamino-but-2-ynyl) 136
2-diethylaminoethylphenylcyclopentane-1-carboxylate, *see* Caramiphen
Diethyl-S-diethylaminoethyl phosphorothiolate, *see* Amiton
1, 5-di (4-ethyldimethylammoniumphenyl) pertan-3-one-diioxide
—structure 73
—use 73
0, 0-diethyl-S-ethylmercaptoethanol thiophosphate, *see* Demeton-S
Diethyl p-nitrophenylphosphate, *see* Paraoxon
0, 0-diethyl-0-(4-nitrophenyl) phosphorothioate, *see* Parathion
Diethyl phosphorofluoridates 5
Diethylphosphoryl-acetylcholinesterase 77
Dihydro-β-erythroidine
—action 190
Diisonitrosoacetone
—structure 78
Diisopropylfluorophosphate, *see* Dyflos
Diisopropylphosphoryl-acetylcholinesterase 77
Dimefox
—poisoning 82
—structure 82
Dimenhydrinate 98
Dimethylaminobornyl acetates 44
2-dimethylaminoethanol, *see* Deanol
3-(2-dimethylaminoethyl) phenyl-N-methylcarbamate
—use 76
Dimethyl-1-methyl-2-carbomethoxy-vinyl phosphate, *see* Phosdrin
Dimethylphenyl piperazinium iodide
—effects 194
—effect on EEG 205
—structure 194
Dimethyl phosphoroflouridates 5
DINA, *see* Diisonitrosoacetone
1, 3-dioxolan
—as muscarinic agent 22, 33
—structure 22
—use in elucidation of antiacetylcholine activity 35
Diphenhydramine
—inhibition of catecholamine uptake 139
—structure 139
2, 2-diphenyl-1, 3-dioxolan 33
Dipterex
—poisoning 114
Di-Systox
—behavioural effects 110

Ditran
—effects on behaviour 160, 169–170, 174, 202
—effects on EEG 149, 152
—effects in man 147
—effects on movement 163–64
—effects on muscle 48
—structure 92
—use 91, 149
DMPP, *see* Dimethylphenylpiperazinium iodide
Dopamine
—uptake 140
Dose ratio 31
Drug action 3, 19
—and behaviour 214–215
Drug
—interaction with receptor 2, 22, 33
DSDP, *see* Amiton
Dyflos
—as anticholinesterase drug 46, 65, 79
—behavioural effects 98, 100, 102, 105, 109–10
—behavioural effects in man 111
—effect on brain acetylcholine levels 136
—effects on EEG 13, 111
at neuromuscular junction 69
—structure 13, 65
—toxicity 70, 73, 77, 82, 85

E

E600, *see* Paraoxon
EA-1701
—toxic effects 112
Edrophonium 72
—structure 65
Eldrin
—poisoning by 114
Enzyme protection in anticholinesterase poisoning 72–76
Enzyme reactivation in anticholinesterase poisoning 76–87
Ephedrine
—structure 94
—use 94
EPN, *see* ethyl-p-nitro-penylthiobenzene phosphonate
EPSP, *see* excitatory postsynaptic potential
Ethyl-N, N-dimethyl phosphoramidocyanidate, *see* Tabun
O-ethyl-S-(2-(N-methyl-N-phenylamino) ethyl) methylthiophosphonate
—structure 84
—toxicity 84
Ethyl-p-nitro-phenyl-thiobenzene phosphonate
—poisoning 94
—structure 94
N-ethyl-3-piperidyl phenylcyclopentyl glycollate
—effect on behaviour 169–70
—effect on movement 163–64
N-ethyl-2-pyrrolidylmethylcyclopentylphenyl glycollate
—effect on behaviour 169–70
—effect on movement 163–64
—effect on muscle 48–49
—structure 48
—use 91, 93
N-ethyl-3-piperidylbenzilate
—effect on behaviour 169–70
—effect on movement 163–64
—tolerance to 151
Excitatory postsynaptic potential 45

F

Fencamfamin
—as muscarinic antagonist 141
"Five-atom rule" 20–22
Fluorine
—liberation of 71
Furan ring 22, 23

G

G3063 91
Gallamine
—structure 7, 88
—use 88
Ganglia, autonomic
—drug action at 44
Ganglion blockade 43
Ganglion blocking agents
—assay 43
—therapeutic uses 87–43
Ganglion stimulation 43, 45–47
Ganglionic receptors 50–54
Ginger Jake paralysis 70
Glycollic acid esters
—antiacetylcholine activity 38, 145
—effect on behaviour 160
—effect on EEG 159
—effect on movement 162
—effect on muscle 50, 144
—poisoning 147, 148
—structure 147

H

H-6, *see* toxogonin
Haloperidol
  —as muscarinic antagonist 141
Hemicholinium 137, 215
  —central effects 195
  —effect on brain levels of acetylcholine 135
  —effects on EEG 206
  —structure 4
Hexamethonium 9
  —action 190, 195
  —action on postsynaptic potentials 45
  —ganglion-blocking activity 51, 87
  —structure 7, 51
  —use 87
Histamine
  —reduction in heart rate by 2
Homatropine
  toxicity 147
Huckel orbital theory 29, 57
Hydrocarbons, chlorinated
  —poisoning 119
Hydrolytic decomposition
  —of acetylcholine 1
Hydrolytic reactivation 65
Hydrolytic reactivation 65
Hydroxamic acids
  —use 77, 79
Hydroxylamine
  —use 77
5-hydroxytryptamine
  —effect on EEG 205
Hyoscine
  —behavioural effects 102, 105, 160, 165–179
  —effect on EEG 152, 155, 157, 160
  effect on movement 162
  —effect on muscle 48
  —isomer of atropine 32
  —as muscarinic antagonist 142, 143
  —poisoning by 11, 147
  —structure 8
  —tolerance to 151
  —use 91
Hyoscyamine
  —effect on behaviour 166, 173
  —effect on movement 162
  —as isomer of atropine 32
Hypoactivity
  —induced by tremorine 141
Hypothermia
  —induced by tremorine 138–141

I

Imipramine
  —as muscarinic antagonist 138, 141, 150
"Induced fit" hypothesis 24
Inhibitory postsynaptic potential 45
  —and α-Adrenoceptor blocking agents 45
Insecticides 5, 56, 71, 97, 111, 113, 115, 118
Intrinsic activity 3
Iproniazid 113
IPSP, *see* Inhibitory postsynaptic potential
Irreversible inhibition 63
Isonitrosodiethylketone
  —structure 80
  —use 80

L

L-dopa 140
Lobelia inflata 12
Lobeline
  —structure 12
LSD-25
  —behavioural effects 202
  —effect on EEG 208
  —tolerance to 151
  —use 149

M

Malathion
  —poisoning 95, 114
Mecamylamine 12, 142, 143, 200
  —effect on EEG 206
  —structure 9
Mercurials, organic
  —poisoning 119
Methacholine
  —structure 19
  —therapeutic uses 19
Methantheline 133
β-methylcholine carbamate, *see* Bethanechol
Methyl demeton
  —behavioural effects 101
  —structure 99
Methylhyoscine 142, 143, 162
N-methyl 3-isopropyl phenyl carbamate
  —behavioural effects 103
  —structure 103

2-methyl-3-oximino-4-oxopentane
—structure 80
—use 80
0-(2-(N-methyl-N-phenylamino)ethyl) methylfluorophosphonate
—structure 84
—toxicity 84, 88
N-methyl-3-piperidyl benzilate
—effect on behaviour 169–70
N-methyl-3-piperidyl benzilate
—effect on EEG 157
N-methyl-3-piperidyl benzilate
—effect on movement 163–64
N-methyl-3-piperidyl benzilate
—structure 145
N-methyl-3-piperidyl phenylcyclopentyl glycollate
—effect on behaviour 169–70
—effect on movement 163–64
MINA, *see* monoisonitrosoacetone
Mipafox
—action 136
—structure 70
Molecular orbital theory 57
Monoisonitrosoacetone
—structure 78
—use 80–82, 85
Morphine 110, 137
Muscarine
—action 190
—enantiomers 24
—structure 6, 22
Muscarinic action 19
Muscarinic activity 20–30
Muscarinic agonists 10, 19–20
—analgesic activity 141
—assay 19
Muscarinic antagonists
—assay 31
—effect on hypothermia 141
—effect on tremor 141
—receptors 34
—stereochemical requirements 33
Muscarinic mechanisms
—in behaviour 176–180
Muscarinic "pharmacophore" 30
Muscarinic receptors, central 132–137
Muscarinic receptors, peripheral 20–30
Muscarinic receptors
—proposed structure 23, 28
Muscarone
—action 190
—enantiomers 24
—structure 22, 23

N

Naphthaquinone
poisoning 119
Neostigmine
—behavioural effects 102, 105, 112
—effect on muscle 68
—effect on respiration 13
—inhibition of acetylcholinesterase 64, 74
—structure 65
—use 5, 63, 73, 76
Nerve gas
—poisoning 116
Neuromuscular blockade 68–69
Neuromuscular blocking drugs 44, 196
—depolarizing 7, 44
—non-depolarizing 7, 44, 53, 88
Neuromuscular junction
—drug action at 7, 44, 50–54
Neuromuscular junction receptors 50, 53
Nicotinamide 94
Nicotine
—action 12, 46, 55
anticholinergic activity 1
—central effects 190–210
effect on behaviour 143, 197–254
effect on brain amines 194–195
effect on EEG 204–209
—enantiomers 56
—structure 6, 56, 191
—toxicity 56
Nicotine-hydroxamic acid methiodide
—use 77
—structure 77
Nicotine monomethiodide
—effect on EEG 205
Nicotinic activity
—of cyclic amidines 54, 56
—molecular requirements 54
—of pyridylpyrrolidines 56
—of tetrahydropyrimidines 54
Nicotinic agonists
—central action 12, 190
—peripheral action 43–45
Nicotinic antagonists 12
—central action 190
—peripheral action 43
Nicotinic receptors 45
—at autonomic ganglia 52–54
—central 190–193
—model of 54–58
—at neuromuscular junction 52–54
Nikethamide
—structure 94
—use 93

p-nitrophenyl diethyl thiophosphate, *see* parathion
Nornicotine
—central effects 193
—structure 194
Nortriptyline
—use 150

O

Occupancy theory 3
Octamethyl pyrophosphortetramide, *see* Schradan
OMPA, *see* Schradan
Onium compounds, synthetic 6
Ordeal bean, *see* Calabar
Organophosphorus agents
—action 5, 69
—effect on EEG 117
—poisoning by 70–72, 79, 87, 94, 113, 119
—use 63
Organophosphorus inhibitors 65, 72, 74
Organosulphates 65
Orphenadrine
—inhibition of catecholamine uptake 139
Ouabain
—reduction in heart rate by 2
Oxamides
—use 74
Oximes
—action on neuromuscular junction 85
—combination with antiacetylcholine drugs 97
—use 70, 77, 79–87
Oxo-oximes
—use 78
2-oxo-3-oximinopentane
—structure 80
—use 80
Oxotremorine
—effect on brain enzymes 136
—effect on brain levels of acetylcholine 135
—inducement of tremor 133, 138
—as a muscarine agonist 28–30, 92
—muscarinic activity 24–26, 133–137
—structure 11, 29, 30

P

2-PAM, *see* Pralidoxime
Paraoxon
—structure 82
—poisoning 82
Parathion
—effect on EEG 117
—poisoning by 73, 95, 96, 114
—structure 96
Parkinsonism, treatment of 32, 137–140
Pempidine 12
—structure 9
Pentamethonium 9
—structure 7
Pentylenetetrazol
—structure 94
—use 94
Pentyltrimethylammonium salt 21
Phenacetylurea
—as nicotinic antagonist 196
Phenindamine
—inhibition of catecholamine uptake 139
—structure 140
Phenobarbital
—as nicotinic antagonist 196
Phenothiazines
—use 91, 95
Phosdrin
—structure 96
—poisoning 96
Phosphorylthiocholines
—toxicity 84
Physostigma venenosum 5
Physostigmine
—anticholinesterase activity 86
—behavioural effects 99, 102, 105, 106, 143, 165, 166, 168, 198, 199
—central effects 195
—effect on acetylcholine levels 135
effect on EEG 13, 159, 207
—effect on muscle 68
—effect on respiration 13
—poisoning 89
—structure 5
—use 73, 74
Pilocarpine
—behavioural effects 110
—structure 10
Pinacolyl-methylphosphonofluoridate, *see* Soman
Piperidyl benzilates
—structure 11
PMCG, *see* N-Ethyl-2-pyrrolidylmethyl-cyclopentylphenyl glycollate
Polymethylene bis-trimethylammonium salts
—ganglion blocking activity 51
Post-ganglionic parasympathetic endings 1, 8

Pralidoxime
—anticholinesterase activity 85
—effect on neuromuscular junction 85
—structure 77
—toxicity 83
—use 77–79, 81, 83–88, 114
Procaine
—action 47, 50, 137
—as nicotinic antagonist 196
—use 87
Promazine
—potentiation of organophosphates 96
Propantheline
—use 87
Propionamides
—nicotinic activity 55
Propionates
—nicotinic activity 55
P2S, *see* pyridine-2-aldoxime methyyl methane sulphonate
Psilocybin
—effect on EEG 208
pyridine-2-aldoxime methiodide, *see* Pralidoxime
Pyridine-2-aldoxime methyl methane sulphonate 78
—use 75, 77, 79, 85, 91, 95, 97
Pyridostigmine
—use 5
Pyridoxine 113
N (pyridyl) methylpyrrolidones
—nicotinic activity 56
Pyridylpiperidines
—nicotinic activity 56
Pyridylpyrrolidines
—nicotinic activity 56

Q

Quaternary ammonium salts 11, 12, 28, 133
—use 74, 93
Quaternary pyridine aldoximes 80
Quinuclidine benzilate
—behavioural effects 160

R

Rate theory 3
Receptors
—interaction with drugs 2, 22, 33
Renshaw cells 9, 190
Respiratory depression 93
Reversible inhibition 63, 72

S

Sarin
—cholinesterase inhibition 78, 80
—structure 70
—toxicity 69–70, 80, 85, 87, 88, 91, 93, 95
Schradan
—behavioural effects 101
Sodium pentobarbital
—effect on EEG 208
—use 95
Sodium perchlorate
—effect on EEG 207
Soman
—poisoning 74, 86
Systox, *see* Demeton-S

T

Tabun
—toxicity 74, 91, 95
TEA, *see* tetraethylammonium
TEPP, *see* tetraethylpyrophosphate
Tertiary amine muscarinic agents 11, 28
Tertiary amines substituted 133
—use 90, 93
Tetraethylammonium 4
—action 190
—structure 8
Tetraethylpyrophosphate
—cholinesterase inhibition 5, 77
—poisoning 82, 90, 116
—structure 6, 77, 82
Tetrahydropyrimidines
—nicotinic activity 54
Tetramethylammonium
—action 46
—structure 8
Tetramethylammonium bromide
—effects 194
—structure 194
Tetrodotoxin 137
Thiohydrocarbons, chlorinated
—poisoning 119
Thyroxine
—effect on EEG 207
TMA, *see* tetramethylammonium
TMB-4, *see* 1, 3-bis-(pyridinium-4-aldoxime) propane dibromide
Tobacco
—central effects 203–204
—effects on EEG 209
TOCP, *see* Sarin
Toxaphene
—poisoning 114

Toxogonin
—structure 79
—toxicity 83
—use 78, 83
Tremorine
—effect on brain enzymes 136
—inducement of tremor 138
—muscarinic activity 24, 25, 135
—structure 11
Triarylphosphates
—poisoning by 70, 71
Triflupromazine
—use 95
—structure 95
Trihexyphenidyl
—effect on movement 162
—inhibition of catecholamine uptake 139
—structure 139
Trimethidinium 143, 200
Trimethylamine
—action 192
Trimethylammonium salts
—structure 21, 22
Triorthocresylphosphate, *see* Sarin
Tropylscopine, *see* Ayoscine
Tropyltropine, *see* Hyoscyamine
Tubocurarine 12
—behavioural effects 111, 112
—at neuromuscular junction 47, 48, 69
—structure 7, 88
—use 88

V

Vagus
—stimulation 1, 2
Vagusstoff 2
Vinyl phosporus esters, halo substituted
—behavioural effects 107

W

War agents 5
Wedensky block 49